I0124764

Shaping the Future with Math, Science, and Technology

Solutions and Lesson Plans to Prepare Tomorrow's Innovators

Dennis Adams and Mary Hamm

ROWMAN & LITTLEFIELD EDUCATION
A division of
ROWMAN & LITTLEFIELD PUBLISHERS, INC.
Lanham • New York • Toronto • Plymouth, UK

Published by Rowman & Littlefield Education
A division of Rowman & Littlefield Publishers, Inc.
A wholly owned subsidiary of
The Rowman & Littlefield Publishing Group, Inc.
4501 Forbes Boulevard, Suite 200, Lanham, Maryland 20706
http://www.rowmaneducation.com

Estover Road, Plymouth PL6 7PY, United Kingdom

Copyright © 2011 by Dennis Adams and Mary Hamm

All rights reserved. No part of this book may be reproduced in any form or by any electronic or mechanical means, including information storage and retrieval systems, without written permission from the publisher, except by a reviewer who may quote passages in a review.

British Library Cataloguing in Publication Information Available

Library of Congress Cataloging-in-Publication Data

Adams, Dennis M.
   Shaping the future with math, science, and technology : solutions and lesson plans to prepare tomorrow's innovators / Dennis Adams and Mary Hamm.
      p. cm.
   Includes bibliographical references.
   ISBN 978-1-61048-116-8 (cloth : alk. paper) — ISBN 978-1-61048-117-5 (pbk. : alk. paper) — ISBN 978-1-61048-118-2 (electronic)
   1. Creative ability in science. 2. Creative ability in technology. 3. Mathematics—Study and teaching. 4. Science—Study and teaching. 5. Technology—Study and teaching. 6. Creative teaching. I. Hamm, Mary. II. Title.
   Q172.5.C74A33 2011
   507.1—dc22

                                                                    2010049971

™ The paper used in this publication meets the minimum requirements of American National Standard for Information Sciences—Permanence of Paper for Printed Library Materials, ANSI/NISO Z39.48-1992.

Printed in the United States of America

Contents

Shaping the Future with Math, Science, and Technology examines how ingenuity, creativity, and teamwork skills are part of an intellectual toolbox associated with math, science, and technology. The book provides new ideas, proven processes, practical tools, and examples useful to educators who want to encourage students to solve problems and express themselves in imaginative ways.

A shallow understanding of math, science, and technology is emblematic of why American students lag behind their counterparts in other countries. The good news is that at least some of our schools are doing a good job of helping their students develop abilities in these subjects. Also, there are teachers who are doing a great job of teaching skills related to imaginative problem solving and innovation (key factors in the evolving global economy).

The development of a technological knowledge-based economy depends on the development of educational systems that allow *all* students, teachers, and schools to do better with both content and imaginative problem solving. And yes, it really is possible to educate our way to a better economy and a better future. Paying attention to twenty-first-century approaches and skills can help accomplish those goals.

Differentiated instruction and collaborative inquiry are two examples of approaches that teachers can use to build on the needs of individual students *and* the social nature of learning. When done right, these and other relatively new instructional methods can spark student interest in mathematical problem solving, scientific inquiry, and technology education.

It is one thing to think about learning possibilities; it is quite another to actually get things done. Like more than a few teachers, some learners simply don't like math and science; too many don't think that they can be successful with these subjects. (Technology has a more attractive image.) At any level, it seems that poor attitude and poor achievement amplify one another.

Changing from *bad attitude, how to get it, and how to keep it* into something more positive is critical if we are going to develop good learning attitudes toward "hard" subjects. Teachers are the key, but real success for learners also has a lot to do with high-quality schools, a supportive home environment, and student self-discipline. Differentiating the learning process can help because it encourages teachers to develop lessons that meet the readiness, interests, and learning profiles of their students.

When it comes to improving teaching, part of the problem is figuring out what makes teachers great and passing on their exemplary skills to colleagues. Whether it is teachers or students, it's hard to be creative if you don't have a clue about the subject. As far as curriculum and instruction are concerned, a key question is whether or not the thought processes and logic being propagated during the teaching of content help *all* students learn how to deal with a rapidly changing environment.

Innovative ability has a lot to do with learning how to synthesize different and often unrelated elements into something new. As far as digital technology is concerned, "open innovation" is favored by some technology companies. This approach is a little like evaluating all the clever ideas in a high-tech suggestion box and coming up with a collective answer. At the opposite end of the spectrum, innovation involves a highly skilled team implementing the creative insights of a powerful leader.

In a society where a glut of information is readily available, it is important to be able to figure out what is important—and do it in a way that allows for building a unique and compelling narrative. It is equally important to learn how to go beyond conventional wisdom and find the truth. As far as the classroom is concerned, building competencies in these areas requires identifying individual strengths, maximizing potential, and building capacity.

There are many compelling reasons for making sure that the humanities and the social sciences are central players in arranging the contours of the future. Even new models of technological innovation often have a lot to do with integrating technological invention into a social context. Understanding technology requires a broad understanding of human thought and behavior. Although the focus here is on the subjects of math, science, and technology, we recognize that they alone can't solve all the problems found in the world today.

When the latest technology is fused with the reasoned development of new knowledge, it can help cultivate a culture that supports creativity in all students. Among other things, it helps if teachers are able to provide support structures and lessons that help everyone in the classroom become better at holding a subject or problem in mind long enough to see it anew. (This is not easy in today's fast-paced multitasking world.)

When students are provided an opportunity to be bold and actually accomplish something, new possibilities open up for them to generate new ideas and collaboratively apply them. This involves balancing good judgment with risk taking and daring to ask thoughtful questions. In this context, having the "right" answer is less important than asking great questions that lead to imaginative possibilities.

Clearly, fostering learners' imaginative skills is of ever increasing importance in our technological society. In tomorrow's schools, an important curriculum goal is bound to involve the application of intellectual tools and technological products of math and science in a way that ignites innovative behavior. The same can be said for preparing students in a way that helps them imagine new scenarios and prepare for future challenges.

What about the nature of the future?

No matter how things turn out, you can be sure that what happens on the road out to (and beyond) the horizon is subject to sudden and unexpected changes. It is also certain that some of today's facts will turn into tomorrow's fiction—and vice versa.

Predicting or controlling the future may be out of the question, but being able to shape it is within the realm of possibility.

CHAPTER 1

Innovation and Differentiation

Twenty-First-Century Math, Science, and Technology Skills

If our schools can ignite sparks of student creativity in association with math, science, and technology lessons, the result is likely to release a mix of innovative ideas and content understanding. Such a combination of imaginative twenty-first-century skills and related intellectual tools is the key to shaping a productive future.

Knowledge of the past can help, but what has already happened is more like a static map that stops at the present. It's not a compass pointing in the right direction. Still, even though events tend to trump trends, we can extend the best ideas available today in a way that gives us some tantalizing hints of future possibilities.

As far as quality schooling is concerned, chance and surprise events have always favored the prepared mind. Now, as in the past, academic success depends on a quality curriculum, student commitment, and highly skilled teachers. How do we recruit and retain good teachers? One way is to look at the competition. In the most academically successful countries, teachers are recruited from among the best college graduates, trained well, paid well, and highly respected.

What about curriculum construction? It begins by determining what we want students to know and be able to do. Some American students do well no matter what. The challenge is making sure that *all* students are motivated and engaged in deep learning.

Treating learners as uniformly passive recipients of information simply won't get the job done in today's environment. Getting everybody willing and able to work up to high standards requires recognition of the fact that learning is more natural and effective when students are not served a one-size-fits-all curriculum.

Differentiated instruction (DI), standards-based lessons, and innovative behavior are compatible with the reality of up-to-date work in math, science, and technology. DI is built on the belief that students have different needs and that every student in the classroom can imaginatively approach any subject.

PROMOTING INNOVATION IN A DIFFERENTIATED CLASSROOM ENVIRONMENT

To differentiate instruction is to recognize students' varying backgrounds, knowledge, readiness, and learning preferences. The next step is making sure that there is some variation in content, process, and/or product. By *content* we mean multiple paths for taking in information; *process* is viewed as multiple options for making sense of ideas; and *product* looks at multiple options for expressing what has been learned. As this structure suggests, differentiated math, science, and technology lessons can be designed in a way that nurtures the innovative spirit.

A differentiated approach to instruction is guided by the idea that schools should extend individual potential and help students to grow as much and as quickly as possible. Unlike some individualized learning approaches, differentiated instruction often happens in small groups with similar interests.

The diversity of thinking within small groups can amplify the innovative ideas generated by individual students. As students collaborate, they can take more responsibility for what they are doing. And they can help each other master content lessons based on learning goals and curriculum standards.

The advance of cognitive learning theories has led educators to realize how important it is for learners to be more actively engaged in the construction of knowledge. As far as innovation itself is concerned, the process begins with creating new ideas in the right environment. Students' attitudes and beliefs are crucial.

Creativity and innovation can build on questioning, trusting, taking risks, being open to new ideas, and having patience. Being able to ask questions that haven't been asked before certainly helps, so does recognizing the fact that innovation can be radical *or* incremental changes in thinking, in processes, in technology, or in services. Changes in thinking, in processes, in technology, or in services involving innovation can be either radical or incremental. Along the way, it always helps if there are time and space for reflection, experimentation, and the collaborative generation of new ideas.

Problem solving in math and inquiry in science are good examples of creative processes by which questions are asked, evidence is gathered, and investigations are carried out. When it comes to technology, the basic idea is to make the world more creative, expressive, empathetic, and interesting. Technology education goals include improving the human condition—not escaping it.

Along with developing explanations and making predictions, innovative skills can help students generate a set of interrelated processes that they can use to investigate phenomena and imaginatively think about real-world problems.

By using digital tools and taking part in mathematical and scientific problem solving, learners can acquire knowledge and develop an understanding of new concepts, models, and theories. For students to understand any academic field, they must learn about the purposes, methods, and thoughtful behaviors associated with these subjects. This implies frequent opportunities to inquire into their own understandings, while learning how to think and work in a way that mathematicians, scientists, and technology experts do.

Understanding interrelated concepts, interpreting explanations, and evaluating evidence are all part of nurturing the innovative process. Helping students become creative and innovative problem solvers is now viewed as an important goal of mathematics, science, and technology instruction. Engaging in tasks before the method of solution becomes clear often results in the development of new knowledge.

The world is an interdisciplinary set of problems out there to explore. Solving a problem often requires finding a way around an obstacle or getting past some difficulty to experience the joy of discovery. To become skillful at this, students need frequent opportunities to formulate, work with, and solve complex problems. The process requires a high degree of student effort and time to allow students to reflect on their own thinking.

Inquiry might be thought of as the use of the processes of science, scientific knowledge, and attitudes to reason and think critically. It is an approach that uses scientific, mathematical, and technological tools to study the natural world and communicate the results to others.

Problem solving is at the core of math instruction—and inquiry plays the same role in science. Both motivate learners and help them understand the nature of math, science, and technology and how these subjects impact their lives (Bass, Contant, & Carin, 2008).

Problem solving, inquiry, and innovative reasoning are natural allies. All involve questioning conventional wisdom and seeking out evidence. We must all recognize that insights are more than personal opinion; points of view must be backed up with evidence and good reasons. It's not just using data and information to support a facts-based argument. It's also important to go beyond the discovery of new facts and *discover new ways to think about them.*

PUTTING THE INNOVATION PIECES TOGETHER

Innovation has a lot to do with the creation of new ideas, approaches, or services. It might also involve transforming an idea or invention into a problem-solving device, process, product, or technique. Uniqueness and novelty are frequent associates.

Intellectual curiosity often inspires the discovery potential and possibilities that others have overlooked. Also, having the curiosity to ask why something works and how something can be done in a different way is a constant part of the process.

Questioning provides room to explore and discover new ideas. Questions can be inquisitive or judgmental. They can convey interest and help students take imaginative steps through investigation and problem solving. The good things that stem from generating good questions can matter more than the answers. Also, even after coming up with informed answers, students can continue to question their assumptions (Browne & Keeley, 2000).

Trust is another piece of the innovation puzzle. At least some of the time, teachers need to trust their students. And students need to trust themselves, other students, and the teacher. This is more likely to happen when students understand that the teacher is there to support them.

Teachers honor student voices by inviting student ideas, encouraging student thoughts, affirming, supporting, and responding with honesty. The world involves challenge and response. When it comes to creativity and innovation, accountability matters and blame doesn't.

Risk is an important part of the innovation picture. It means taking informed chances. An exciting element of risk: *is it going to work, or not?* The eagerness to take bold (yet calculated) risks is one of the basic principles of innovation. The possibility of failure goes hand in hand with risk taking. Just try to learn from mistakes. Accepting failure is not easy, and often painful. You have to know when to hold them and *when to fold them.* Being able to give up what we passionately believe requires tolerance and patience.

BEING OPEN TO NEW IDEAS

Being open to new ideas is another basic element of innovation. It requires an open mind and atmosphere that encourages people to imagine, think broadly, and collaborate. When it comes to innovation, experience can be both a benefit and a handicap. Being open to new approaches can capture serendipity and give those involved the freedom to create. Part of openness involves curiosity related to the ability to critically evaluate data, accept input, and be ready to adapt to change.

At times it is lack of imagination that kills creative work and stifles innovation—at other times, it is the perceived need that full agreement must be reached. Being able to question rituals is an important step toward realizing that just because something has always been done a certain way doesn't mean it can't be changed for the better.

When groups are working on new ideas or approaches, everyone should be encouraged to speak freely; but it is also important to realize that consensus may not be possible. Once a path is chosen, it is up to everybody to open their minds, focus, and execute the plan. Dealing with surprises and sharing information create avenues for valuable feedback.

A certain level of *patience* is mandatory if innovation is to succeed. Being patient allows ideas to ripen. Successful innovators tend to be better at enduring uncertainty, to doubt and wait, instead of jumping at the first solution that comes along. Some new things turn up by accident, like Fleming's discovery of penicillin. At other times, break-throughs are made simply because those involved didn't know any better, like Watson and Crick's discovery of the DNA molecular structure (the double helix). Still, in both cases, those involved were first-rate scientists and could recognize their discoveries.

Overcoming obstacles and defending daring ideas are part of the creative and innovation process. Individuals who are highly innovative also seem to be able to cut through the clutter of information and argument to locate gems of wisdom. Although some great innovations have come from exceptional insights, many have a lot to do with nurturing unique perceptions. Also, readiness for opportunities and the element of chance play major roles (Zmuda, 2010).

When it comes to debating various issues, learners need to realize that the goal isn't necessarily displaying and defending the superiority of your own view. Sometimes, the most important thing is taking what you believe and turning it into something at least a little different and better.

Teachers who strive to promote creativity and innovation know that one size never fits all. Differentiated instruction can be a helpful partner; it accommodates different student needs and interests. Differentiated lessons frequently involve small group instruction, collaborative projects, personalized suggestions, open discussions, and other strategies to make learning meaningful for each student. Experienced instructors also tend to use a mixture of student-choice and teacher-choice assignments.

ACCOMMODATING STUDENT DIFFERENCES

Schools are adapting to an increasingly broad range of learning styles and academic competencies. So it is more important than ever to design curriculum and instruction in a way that works for academically diverse learners. Differentiated instruction is a natural partner in such an accommodation.

With a standards-based curriculum in place in almost every state, all students are now expected to achieve at high levels. In the past, there was a group of students who ended up being placed in low-level classes. Now we teachers are expected to differentiate in ways that ensure that everyone does some higher-level thinking.

To differentiate instruction, teachers know the needs of their students and how to make learning personally relevant for them. It also requires that students' academic growth be closely monitored throughout the learning cycle. This requires authentic assessment before, *during*, and after instruction.

Armed with good information and knowledge of their students, teachers can adapt their teaching plans to accommodate varied levels of readiness, interests, and learning styles. Experienced teachers know that when learners are interested in the topic—and know they can do related instructional tasks—they are more likely do whatever it takes to succeed.

Comprehending what is being read is crucial. Part of the answer: reading partnerships, text at different reading levels, independent studies, varied homework assignments, and personalized evaluations. Portfolios can help; they require students to collect, select samples, and reflect on their work. Another approach is to use ongoing formative assessment practices connecting the latest information to curriculum and instruction. Sometimes, simply observing students as they work gives teachers clues for adjusting lessons and assignments.

Letting students make responsible decisions concerning how to use some of their time is one way to differentiate and let students exercise some power over their learning. Also, teachers can respect student ideas by inviting their ideas, encouraging participation, and being honest (Tomlinson, Brimijoin, & Narvaez, 2008).

In designing a differentiated lesson plan, teachers make time for student discussions, encourage problem solving in small groups, and ask for student input in developing classroom activities. Differentiated feedback from the teacher can involve just rephrasing a question, for instance—or regrouping on the basis of student needs and interests.

DIFFERENTIATED LEARNING GOALS

Differentiated instruction puts students as close as possible to the center of learning, while providing space for meeting learning needs. The first step in differentiating learning is to begin where you are. Differentiation can be built into lessons from past years by providing ways to gear up and gear down assignments. The process also involves analyzing how well you're providing variety and challenge in learning, recognizing which students are best served by your current plans, and altering those plans as needed so more students can be successful.

Differentiated learning helps students not only master content but also shape their learning identities (Chapman & King, 2009). Differentiated instruction increases learning for all students by involving them in activities that respond to their individual strengths and inclinations. When it comes to math, science, and technology, differentiated instruction goals include

- Developing challenging and engaging tasks for every student.
- Creating instructional activities founded on the important topics, concepts, and skills.
- Providing ways for students to display what they have learned.
- Offering flexible approaches to content and instruction.
- Paying attention to students' readiness, instructional needs, and learning preferences.
- Meeting curriculum standards for each learner.

Differentiated learning involves adapting instruction to meet the needs of students with a wide range of needs and academic abilities. It also involves recognizing the fact that individuals and small clusters of students can use different content, processes, and products to achieve the same conceptual understanding.

THE BASIC ELEMENTS OF DIFFERENTIATED INSTRUCTION

The teaching *content* and ways students have access to information are important ways for teachers to differentiate instruction. Teachers usually select content to teach one or more standards. Teaching content is often decided by the school or district and reflects state or national standards. It is important to know your content thoroughly so it will be easier to decide which standards to use.

Required skills are presented to teachers as mandated standards. Administrators often give teachers freedom to select the content. You differentiate the content when you preassess students' skills and knowledge, then match students with activities according to their current knowledge, understanding, and skill level. Readiness reflects what a student knows, understands, and is able to do.

Interest is another important way to differentiate learning. What students enjoy learning about, thinking about, and doing provide a motivating link. Teachers connect required content to students' interests to engage the learner. Students connect with new information by making it appealing, relevant, and worthwhile.

A student's *learning profile* is influenced by his or her preferred learning style, preference, "intelligence," academic interests, and cultural background. By examining a student's learning profile, teachers can extend the ways students learn best.

A *learning environment including differentiation* allows teachers and students to work in ways that benefit each student and the class as a whole. A flexible environment allows students to make decisions about making the classroom surroundings work better. This provides students a feeling of ownership and a sense of responsibility. Students of all ages can work successfully as long as they know what's expected of them and are held to high standards of performance.

IMPORTANT PRINCIPLES OF DIFFERENTIATION

The key principles of differentiated instruction include the following:

1. A *quality, engaging curriculum* is the beginning principle. Your first job as a teacher is to guarantee that the curriculum is consistent, inviting, important, and thoughtful. Students' work should be appealing, inviting, thought provoking, and stimulating. *Each student should find his or her work powerful and interesting.*

2. Teachers *give students challenging tasks* that are a little too difficult. Be sure there is a support system to assist students' success at each level.

3. It's important to *use adjustable grouping.* You need to plan times for groups of students to work together—and times for students to work independently.

4. *Assessment is a continuing and ongoing process.* Teachers often preassess students to determine students' knowledge and skills based on their needs. Once you are aware of what students already know and what they need to learn, then you can differentiate instruction to match

the needs of each student. Formative assessment allows teachers to alter instruction while learning is going on.

When it comes to planning lessons or assignments, we often use a *tiered approach*. Tiered ways of teaching are differentiated learning ways to teach based on your diagnosis of students' needs. When you use a tiered approach, you are prescribing individual techniques to particular groups of students. Within each group, you decide whether students do assignments alone, with a partner, or as a learning team.

The basic idea is to have a wide range of students learn the concept being taught. But students can reach competency in many different ways. The first step is to identify the key skills and concepts that everyone must understand. All students in a class cover the same topic, but the teacher uses different materials based on students' aptitudes and interests.

5. *Some student grades should be based on growth.* A struggling student who persists and doesn't see progress will likely become frustrated if grade-level assignments seem out of his or her reach and growth doesn't appear to count. It is your job to support the student by making sure (one way or another) he or she masters the concepts required.

When it comes to differentiated instruction, teachers need to realize that differentiation is about grouping according to the interests and social, academic, and emotional needs of students. It's not about ability grouping. Teachers need to build motivating lessons for mixed-ability groups. And they need to continuously think about their teaching so that they can quickly adapt their approach as situations change.

As the standards suggest, educators should reflect on their work and pay attention to how theory and research might apply to practice.

MULTIPLE INTELLIGENCES AND DIFFERENTIATED INSTRUCTION

The first step in differentiating instruction is getting to know each student's learning preferences and levels of readiness and motivation. Since students learn and create in different ways, it is often best to teach to students' strengths, providing learners with deep learning experiences in different ways that can enrich their "intelligence" in specific areas.

Descriptions of intelligence have frequently focused on mental abilities like problem solving, comprehension, memory, and the ability to think metaphorically. There are, however, different ways of looking at it. Robert Sternberg (2011) believes that a broader definition is useful. He suggests three intelligence preferences:

Analytic (schoolhouse intelligence): Analytical intelligence is the ability to complete problem-solving tasks.

Creative (imaginative intelligence): Creative ability is viewed as being able to deal with new or unusual situations by drawing on existing knowledge and skills.

Practical (contextual, street smarts): This "intelligence" may be considered the ability to adapt to everyday challenges and understand what needs to be done in a specific situation.

Like other ideas about multiple intelligences, there is very little empirical evidence for or against the basic concept. Still, it seems reasonable to believe that the traditional definition of intelligence does not sufficiently encompass the wide variety of abilities needed for success in school and in life. The idea that there are separate human capacities has powerful implications for teaching and learning.

Howard Gardner suggests that intelligence can be viewed as a variety of independent intelligences rather than an overall measure of mental ability (2006). Although the exact number is not carved in stone, he has identified at least eight intelligences that represent different ways children understand or explain their knowledge. These include

Linguistic: the ability to use language and express ideas.

Logical/mathematical: the ability to examine patterns and relationships by manipulating objects or symbols in an orderly manner.

Musical: the ability to think in music, the capacity to perform, compose, or enjoy a musical piece.

Spatial: the ability to understand and mentally manipulate a form or object in a visual or spatial display.

Bodily-kinesthetic: the ability to move your body through space and use motor skills in sports, performing arts, or art productions (particularly dance or acting).

Interpersonal: the way you work with groups—interacting, sharing, leading, following, and reaching out to others.

Intrapersonal: the ability to understand your inner feelings, dreams, and ideas.

Naturalist: the way you discriminate among living things (e.g., plants, animals) as well as a sensitivity to the natural world.

We should note that Gardner has explored additional possibilities such as existential, spiritual, and moral intelligence.

Although there is still discussion about what these intelligences represent, or if there are additional intelligences, our purpose here is to show a broad collection of instructional activities that offer learning choices for each of the intelligences.

Students have many strengths and weaknesses, and the effective teacher can put into action lessons that provide learning activities that speak to a wide assortment of these intelligences to provide the best learning opportunities for each student. Teachers are encouraged to plan their lesson activities with these intelligences in mind.

Gardner's work can help you add excitement to the ways you teach and the projects you assign. It can also help you find out more about your students, the ways they learn best, and what they like to do. The activities in this book look at student actions using differentiated instruction.

ACTIVITIES BASED ON MULTIPLE INTELLIGENCE THEORY

1. One way to find out students' interests and strengths is to explain multiple intelligence (MI) theory to them and provide them with a list of possible activities. Have them underline the ones they would like to try.

2. Have your students review the MI activities list and note which intelligences had the most activities underlined. Make sure your students understand the purpose of this activity is to find out their learning strengths.

Student Actions: Using Direct Instruction

Read the lists and underline all the MI activities that you enjoy doing, presenting, or performing. Here are some possibilities:

linguistic intelligence
writing about something that
   happened
designing a newscast
forming a plan, describing a
   procedure
sending a letter, writing a play
interpreting text or a piece of writing
conducting an interview, debating

musical intelligence
performing a rap song
giving a musical presentation
explaining music similarities and
   differences
demonstrating rhythmic patterns
performing music

logical/mathematical intelligence
conducting an experiment
explaining patterns
designing analogies to explain
solving a problem
inventing a code
creating a website

spatial intelligence
illustrating, sketching
forming a slide show
creating charts, maps, or graphs
designing a piece of art
drawing, painting
videotaping

bodily/kinesthetic intelligence
using creative movement
designing task or puzzle cards
building or constructing something
using hand-eye coordination
using the body to persuade or
   support others
using technology to explain

interpersonal intelligence
participating in a service
   project
conducting a meeting
teaching someone
consoling others
advising or supporting a friend

naturalist intelligence
preparing an observation notebook
knowing about nature
working with nature
describing changes in the
   environment
caring for pets, gardens, or parks
using binoculars, telescopes, or
   microscopes
photographing natural objects

intrapersonal intelligence
describing one of your values
being independent
feeling comfortable with your
   thoughts
reflecting on emotions
assessing your work
setting and pursuing a goal
dreaming

analytic   *creative*
reviewing basic skills   imagining or creating

contextual
being street smart

3. Once students have expressed their choices, have them perform some activities to help them remember their intelligences.

4. Discuss various learning styles:

Concrete style learner: follows a step-by-step process, learns sequentially.

Abstract style learner: focuses on ideas and abstractions, learns through a process of questioning.

Self-expressive style learner: looks for images, uses feelings and emotions.

Interpersonal style learner: focuses on the concrete, prefers to learn socially and judges learning in terms of its potential use in helping others.

Encourage students to identify their preferred learning style and have them get together with others and review learning preferences.

5. Build on students' interests. When students do research either individually or with a group, allow them to choose a project that appeals to them. Students should also choose the best way for communicating their understanding of the topic. In this way, students discover more about their interests, concerns, learning styles, and intelligences.

6. Plan interesting lessons. There are many ways to plan interesting lessons. Lesson plan ideas presented here are influenced by ideas as diverse as those of John Goodlad, Madeline Hunter, and Howard Gardner.

Lesson Planning

1. Focus student attention and relate the lesson to what students have done before. Stimulate interest.

2. Present the objectives and purpose of the lesson. What are students supposed to learn? Why is it important?

3. Provide background information: what information is available? Resources such as books, journals, videos, pictures, maps, charts, teacher lectures, class discussions, or seat work may be listed.
4. Define procedures: what are students supposed to do? This includes examples and demonstrations as well as working directions.
5. Monitor students' understanding. During the lesson, the teacher may check students' understanding and adjust the lesson if necessary. Teachers invite questions and ask for clarification. A continuous feedback process is always in place.
6. Present guided practice experiences where students have a chance to use the new knowledge presented under direct teacher supervision.
7. Give students many opportunities for independent practice where they can use their new knowledge and skills.
8. Assess and evaluate students' work to show what students have done.
9. Refer to the lesson plan formats included in the resources at the end of the chapter.

A SAMPLE OF MULTIPLE INTELLIGENCE LESSON PLANS

Differentiated Brain Lesson: Discover How Neurons Work

The human body and the brain are fascinating areas of study. The brain, like the rest of the body, is composed of cells; but brain cells are different from other cells. Neurons grow and develop when they are actively used and they diminish when they are not used.

High-interest materials such as colorful charts are desirable when teaching about neurons. A chart of a neuron in the brain including cell body, dendrite, and axon is a helpful teaching tool when introducing this concept. Drawing on the latest research in brain-based learning, differentiated instruction, and multiple intelligences, this lesson provides strategies that motivate students (Tate, 2004).

Lesson goals: A major goal is to provide students with a dynamic experience with each of the eight intelligences and have students map out a chart on construction paper.

Procedures:

1. Assign the class to groups, with each group representing a specific type of intelligence.
2. Give students time to prepare an activity that addresses their assigned intelligence.
3. Each group will present a three-minute presentation (including their chart) to the entire class.
4. Allow students to see, hear, touch, and write about new or difficult concepts.
5. Use materials that assess learner needs. Have the class develop their own problems.

Objective: Introduce students to new vocabulary about how the brain functions, specifically the functions of the neurons.

Grade level: With adaptations, K–8.

Materials: Paper, pens, markers, copy of the picture of the brain, the neuron, songs about the brain, model of the brain.

Background information: Very few, if any, people understand exactly how the brain works. But scientists know the answer lies within the billions of tiny cells called *neurons*, or nerve cells, that make up the brain. All our body's feelings and thoughts are caused by the electrical and chemical signals passing from one neuron to the next. Neurons carry signals throughout the brain that allow the brain to move, hear, see, taste, smell, remember, feel, and think.

Lesson procedures:

1. Show a picture of the brain and a neuron and explain its various parts.
2. Pass out brain pictures.
3. Have the students label the parts of the neuron and color if desired.

Activity 1: Explain How Neurons Work

Linguistic Intelligence (sending message transmissions)   A message traveling in the nervous system of the brain can go 200 miles per hour (mph). These signals are transmitted from neuron to neuron across synapses (a point at which a nervous impulse passes from one neuron to another) (Walker Tileston, 2010; Willingham, 2010).

1. Instruct students to get into groups of five. Each group chooses a group leader.
2. Direct students to stand up and form a circle. Students should be an arm's length away from the next person.
3. Each person is going to act as a neuron.
4. When the group leader says, "Go," have one person from the group start the signal transmission by striking with an open hand the hand of the adjacent person. The second person, then, strikes the hand of the next, and so on, until the signal goes all the way around the circle and the transmission is complete. Use materials that address learner's needs. Allow the entire class to become involved in this learning activity. This helps all students realize they can work successfully and have fun with peers.

Explanation: The hand that receives the open hand strike (or slap) is the "dendrite." The middle part of the student's body is the "cell body." The arm that gives the slap to the next person is the "axon," and the hand that gives the slap is the "nerve terminal." In between the hands of two people is the "synapse."

Inquiry questions: As the activity progresses, innovative questions will arise such as "What are the parts of a neuron?" A neuron is a tiny nerve cell, one of billions that make up the brain. A neuron has three basic parts—*the cell body, the dendrites, and the axon.* Have students make a simple model by using their hand and spreading their fingers wide. The hand represents the cell body, the fingers represent dendrites, which bring information to the cell body, and the arm represents the axon, which takes information away from the cell body. Just as students wiggle their fingers, the dendrites are constantly moving as they seek information.

If the neuron needs to send a message to another neuron, the message is sent out through the axon. The wrist and forearm of the student represent the axon. When a neuron sends information down its axon to communicate with another neuron, it never actually touches the other neuron. The message goes from the axon of the sending neuron by "swimming" through the space called the synapse. Neuroscientists define *learning* as *two neurons communicating with each other.* They say neurons have "learned" when one neuron successfully sends a message to another neuron.

Activity 2: Join or Link the Dots

Spatial Intelligence and Logical/Mathematical Intelligence *Lesson goals*: This activity is designed to show the complicated connections of the brain.

Materials: Colored pencils or markers, typing paper.

Procedures:

1. Have students make ten dots on a side of a sheet of typing paper and ten dots on the other side of the paper.
2. Have students imagine these dots represent neurons. Assume each neuron links or makes connections with the ten dots on the other side.
3. Then, have students draw lines to connect each dot on one side with the dots on the other side. This is a simple view of what actually happens. Each neuron (dot) may actually make thousands of connections with other neurons.

Multiple Intelligences Activities for Differentiated Brain Lesson

Musical Intelligence Teaching brain songs to students is another part of this Multiple Intelligence Brain Lesson.

1. Teach a small group of students the words and the melody of the songs "I've Been Working on My Neurons" and "Because I Have a Brain." Include all students in this song process. Students will become the brain song "experts."

> "I've Been Working on My Neurons" (sung to the tune "I've Been Working on the Railroad")
> I've been working on my neurons, all the live long day.
> I've been working on my neurons, just to make my dendrites play.
> Can't you hear the synapse snapping? Impulses bouncing to and fro,
> Can't you tell that I've been learning? See how much I know!

> "Because I Have a Brain" (sung to the tune "If I Only Had a Brain")
> I can flex a muscle tightly, or tap my finger lightly,
> It's because I have a brain,
> I can swim in the river, though it's cold and makes me shiver,
> Just because I have a brain.

I am really fascinated, to be coordinated,
It's because I have a brain.
I can see lots of faces, feel the pain of wearing braces,
Just because I have a brain.
Oh, I appreciate the many things that I can do.
I can taste a chicken stew, or smell perfume, or touch the dew.
I am heavy with emotion, and often have the notion,
That life is never plain.
I have lots of personality, a sense of true reality,
Because I have a brain.

Interpersonal Intelligence
2. The "experts" will teach the songs to the rest of the class.

Activity 3: Introduce Graphic Organizers

Intrapersonal Intelligence   Graphic organizers help students re-member information. Mind mapping or webbing shows the main idea and supporting details. To make a mind map, write an idea or concept in the middle of a sheet of paper. Draw a circle around it. Then, draw a line from the circle. Write a word or phrase to describe the concept; draw other lines coming from the circle. Then have students draw pictures to represent their descriptions. Students can start mapping by examining the skill section of their map. Encourage students to talk about word choices and share their picture creations.

Activity 4: Review and Assess the Lesson

Linguistic, Bodily-Kinesthetic, Spatial, Logical/Mathematical, In-terpersonal, Naturalist, and Intrapersonal Intelligences   The activi-ties on the Differentiated Brain Lesson shown here help students reveal their exciting curiosity as they observe, manipulate, and sort common objects and materials in their environment. Students find musical and mathematical patterns. They continue to explain their ideas of the world as they work with national math and science education standards.

The lesson activities "Explain How Neurons Work," "Join or Link the Dots," singing songs, making mind maps, and reviewing and as-sessing focus on the **math standards** of problem solving, estimation, data analysis, logic, reasoning, communication, and math computations. The **science standards** of life science, inquiry, science communication, science and technology, and personal and social perspectives are also addressed.

INNOVATIVE SCIENCE AND MATH ACTIVITIES

The following activities are designed primarily for elementary and middle school students.

Activity 1: Using the Five Senses to Observe, Describe, and Share Information

Description: This activity uses the five senses. The process skills of observing, inferring, communicating (sharing), and hypothesizing are introduced.

Planning group: Group members should help plan and arrange the classroom and materials.

Objectives:

1. Students will make inferences by observing and using their senses.
2. Students will communicate and share their ideas with others.
3. Students will question and make hypotheses based on their senses.
4. Students will validate their thinking through personal experiences.

Procedures:

1. Select objects that are safe to touch, smell, and taste. Objects such as cookies, oranges, apples, and popcorn are good choices.
2. Place one object in a clean paper bag and have students feel the object without looking inside.
3. Next, have students describe what they feel.
4. Encourage students to smell an object without peeking.
5. Have students talk about what they smell.
6. Shake the bag and tell students to describe what they hear.
7. You may wish to have students taste the object and describe it.
8. Finally, have students look at the object and validate their guesses.

It's important to talk with students about the plans they used in making their guesses. Be sure you discuss the valuable role of their classmates. Ask what they learned from each other about making inferences. Experiences such as these give students an opportunity to develop and refine many science and math concepts. Students may use vague or emotional terms rather than specific descriptive words. Have students discuss which words give better descriptions. Have students relate their everyday language to science and math language and symbols. The following activity gives students a chance to refine their communication skills.

Activity 2: Describe What You See

Description: This activity uses the process skills of observing, inferring, measuring, comparing, and recording. Students work in group centers.
   Objectives:

1. Students will accurately observe and record data.
2. Students will begin using scientific equipment.
3. Students will show how to work in groups in an organized and effective manner.

   Process skills: Comparing, measuring, ordering by distance, inferring, making conclusions.
   Planning group: Group members should arrange the classroom and materials.
   Materials:

a. Five samples: a computer disk, a flower, a housefly, a piece of fabric, and a piece of paper. (Other samples can be substituted for those listed.)
b. Twelve magnifying glasses.
c. Six rulers.
d. An observation sheet for each student.

   Procedure:

1. Six stations are set up around the room, including two magnifying glasses, one of the samples listed above, and a ruler.

2. Each student gets an observation sheet.
3. Students are divided into six groups of four students each.
4. Each group is assigned a station. At their station, each group has ten minutes to record as many observations about the sample as possible.
5. Using the magnifying glass and a ruler, each student in the group makes an observation for the group to record. Students take turns as time allows.
6. As a class, students compare and discuss their observations.
7. Students are actively involved whether they're the group leader or part of the team. If students have difficulty, encourage them to work as a partnership.

Evaluation: Group data sheets are evaluated on organization, observation skills, and accuracy. Also, it is more motivational if students can see the usefulness of what they are learning.

Many of the suggestions we have presented will work without serious disruption in just about any classroom. However, experience has shown that larger changes in educational approaches are most successful where there is a good relationship among teachers, parents, students, and school administrators. Teachers may be the key, but creative collaboration allows for innovative solutions that they cannot accomplish alone. Widespread restructuring is usually easier when there is an arrangement for consultation among the major players.

REACHING EVERY STUDENT WITH DIFFERENTIATED INSTRUCTION

Successful differentiated classrooms are full of energy, excitement, and the possibility of teaching all students no matter what learning style they prefer. So it is little wonder that many educators find differentiated instruction useful in working with students with a variety of academic strengths. With all the positive possibilities, teachers are coming to view differentiated instruction as an important ally in meeting the needs of today's students—who come to school with increasingly diverse cultural backgrounds, levels of prior knowledge, and interests.

Differentiated learning is responsive to specific individual and small group needs—as well as class performance as a whole. By preparing lessons that differentiate learning, teachers can meet the needs of every student in today's diverse classrooms. It has proven to be a solid asset for teachers trying to reach students who are performing at varying levels in science and math. It is an organized, yet flexible, way of adjusting teaching and learning to meet students where they are and help them accomplish more academically.

When teachers provide alternate paths for getting students to understand a concept, it helps students become self-reliant and motivated learners. It's also a good way to meet individual and group needs in the regular classroom. Since students don't all learn at the same rate, it is important to consider the pacing of math, science, and technology instruction when figuring out the differentiated options. Although it may sound contradictory, the reality is that collaborative groups can provide different individual paths for learning math and science.

It doesn't make sense to reward students for every trivial thing or say, "That's good," when it isn't. Paying attention to whatever they have to say is important. Not every idea is good, but when students feel they aren't being listened to, they go silent.

Students' personal interests, learning profiles, and curiosity about a specific topic or skill are major considerations in differentiated problem-solving and inquiry activities. An added advantage to such an approach is that teachers often report that they enjoy teaching more (Sutman, Schmuckler, & Woodfield, 2008).

At the intersection of differentiation and innovation, there is real power in coupling teamwork with individual accountability. By having the opportunity to collaboratively explore ideas, even unmotivated students tend to respond to appropriate challenges and enjoy learning about science and math. Flexible grouping and pacing, tiered assignments, performance assessment, and other factors associated with DI can inject fresh energy into math, science, and technology instruction.

Differentiated instruction is a proven path to principles that can guide instruction and help teachers modify or adapt subject matter content. Experienced teachers also know that it is a good idea to make sure that students have access to more good ideas and problems than they can work on. This leaves more room for more choice and serendipity.

Some of the ideas set aside can inspire students to creatively rethink whatever it is they have been focusing on. Also, encouraging students to bring new ideas forward is more effective if they don't have to worry about premature evaluation.

SUMMARY AND CONCLUSION

Whether it's in or out of school, there are numerous examples of people making mistakes because they have a narrow range of skills (like technical knowledge) and don't understand the social context in which they are working. So it's important not to focus on a narrow definition of talent and not to forget about skills like empathy, teamwork, and being a self-starter.

The advance of cognitive learning theories confirms what many experienced teachers already know—students need to be broadly prepared and actively involved in their own knowledge construction. In life and in school, it is not always what you know but how active you are in fashioning your own learning. Two other important points that have a lot to do with learning twenty-first-century skills are how quickly you can learn new things and how effective you are in making good use of your imagination.

Innovation has a lot to do with enhancing performance in a digital age. We constantly see the results (good and bad) of the innovative spirit all around us. On the mostly negative side, you have financial meltdowns caused by so-called innovative approaches, like unregulated derivatives. So there are times when the word *innovation* is used to cover up bad behavior, with approaches that if they were a book might be entitled *Accounting Fraud for Dummies*.

On the more-or-less positive side, you have what may be the most important innovation of the late twentieth century, the Internet (Wu, 2010). To figure out what's good, bad, and in between, students must be able to ask the right questions and deal with things they can't predict. Being able to deal with ambiguity is, after all, a part of innovative behavior.

As far as the classroom is concerned, creative and innovative skills can be developed directly or integrated into the context of mathematical, scientific, and technology lessons. In the differentiated classroom, students are encouraged by a variety of approaches that ignite connection and conversation. Clearly, moving in the most productive direction requires treating all students as decision makers, active learners, and innovative problem solvers.

The intellectual tools used by mathematicians and scientists as they go about investigating the world also work for students. Problem solving and inquiry can motivate and encourage all types of students to ask thoughtful questions. Innovation itself can be encouraged by making sure that students have the opportunity to express themselves in different ways, with different media. It also means making sure that lessons leave ample room for open-ended problem solving and innovative behavior.

The learning of math, science, and technology is amplified when students comprehend the nature of these subjects and their relevance to their lives. This is helped along when students are able to connect prior knowledge to observations and use evidence to increase their personal knowledge of how the world works. Creativity and innovation can be ignited by encouraging students to express themselves in multiple media and by engaging them in tasks that require collaborative inquiry and creative thinking.

Science and its mathematical/technological tools have a lot to tell us about the future. Literature adds possibilities like helping us understand who we are and how we might deal with sudden and unexpected changes of fortune. You can weigh the available evidence and extrapolate from existing trends; but at its best, thinking about the future is only sophisticated guesswork.

It is important not to succumb to the temptation to try and foretell the future. Still, the future has not been predetermined; it is a place where teachers and their students are headed.

LESSON PLAN APPENDIX

A Lesson Plan Format for Direct Instruction

Topic and Grade Level
Objective
Theme and/or Motivation
Materials:

1.
2.
3.

Launching the Lesson

Whole-Class Teacher Instruction: List the concepts, definitions, and processes to be used in instruction. List the directions for activities and the examples you will use.

Class/Group/Individual Activities: Include several harder and several simpler problems that can help students at all levels of competency learn the same concept. How are you going to provide for different interests, needs, and aptitudes?

Summarize: How will you decide whether students have learned what you wanted?

Lesson Plan Outline for Group Investigations

Topic of lesson and grade level.
What do you want students to learn?
Why are the concepts important?
What background information do students need before starting?
Organization and procedures.
List the materials needed:

1.
2.

How are you going to get the students involved?
Lesson development, questions, and desired product.
Small group options.
Gearing up (if the lesson is too easy).
Gearing down (if the lesson is too hard).
Assessment (observations, products produced, portfolio entry, etc.).

REFERENCES

Bass, J. E., Contant, T. L., & Carin, A. A. (2008). *Activities for Teaching Science as Inquiry*, 7th ed. Boston: Allyn & Bacon.

Browne, M. N., & Keeley, S. M. (2000). *Asking the Right Questions: A Guide to Critical Thinking*, 6th ed. Upper Saddle River, NJ: Prentice Hall.

Chapman, C., & King, R. (2009). *Differentiated Instructional Strategies for Writing in the Content Areas*, 2nd ed. Thousand Oaks, CA: Corwin Press.

Gardner, H. (2006). *Multiple Intelligences: New Horizons in Theory and Practice.* New York: Basic Books.

Sutman, F. X., Schmuckler, J. S., & Woodfield, J. D. (2008). *The Science Quest: Using Inquiry/Discovery to Enhance Student Learning, Grades 7–12.* San Francisco: Jossey-Bass.

Tate, M. (2004). *"Sit & Get" Won't Grow Dendrites: 20 Professional Learning Strategies That Engage the Adult Brain.* Thousand Oaks, CA: Corwin Press.

Tomlinson, C., Brimijoin, K., & Narvaez, L. (2008). *The Differentiated School: Making Revolutionary Changes in Teaching and Learning.* Alexandria, VA: ASCD.

Walker Tileston, D. E. (2010). *Ten Best Teaching Practices: How Brain Research and Learning Styles Define Teaching Competencies.* Thousand Oaks, CA: Corwin Press.

Willingham, D. T. (2010). *Why Don't Students Like School: A Cognitive Scientist Answers Questions about How the Mind Works and What It Means for the Classroom.* San Francisco, CA: Jossey-Bass.

Wu, T. (2010). *The Master Switch: The Rise and Fall of Information Empires.* New York, NY: Alfred A. Knopf.

Zmuda, A. (2010). *Breaking Free from Myths about Teaching and Learning: Innovation as an Engine for Student Success.* Alexandria, VA: ASCD.

RESOURCES

Adams, D., & Hamm, M. (1998). *Collaborative Inquiry in Science, Math, and Technology.* Portsmouth, NH: Heinemann.

Armstrong, S., & Haskins, S. (2010). *A Practical Guide to Tiering Instruction in the Differentiated Classroom: Classroom-Tested Strategies, Management Tools, Assessment Ideas, and More to Help You Create Effective Tiered Lessons That Work for Every Learner.* Scranton, PA: Scholastic Teaching Resources.

Arnold. A. (2010). *Stimulating Creativity and Enquiry (Key Issues).* London: Featherstone Education.

Craft, A. (2010). *Creativity and Education Futures.* Staffordshire, UK: Trentham Books.

D'Amico, J., & Gallaway, K. (2010). *Differentiated Instruction for the Middle School Science Teacher: Activities and Strategies for an Inclusive*

Classroom (Differentiated Instruction for Middle School Teachers). San Francisco: Jossey-Bass.

Dodge, J. (2009). *25 Quick Formative Assessments for a Differentiated Classroom: Easy, Low-Prep Assessments That Help You Pinpoint Students' Needs and Reach All Learners*. Scranton, PA: Scholastic Teaching Resources.

Estrin, J. (2009). *Closing the Innovation Gap: Reigniting the Spark of Creativity in a Global Economy*. New York: McGraw-Hill Companies.

Glass, K. T. (2009). *Lesson Design for Differentiated Instruction, Grades 4–9*. Thousand Oaks, CA: Corwin Press.

Grabosky, D. M. (2009). *Real-World Investigations for Differentiated Instruction*. Columbus, OH: Macmillan/McGraw-Hill/Glencoe.

Hanson, H. M. (2009). *DI: Differentiated Instruction Enhancing Teaching & Learning*. Port Chester, NY: National Professional Resources/Dude Publishing.

———. (2009). *Response to Intervention & Differentiated Instruction*. Port Chester, NY: National Professional Resources/Dude Publishing.

Heacox, D. (2009*). Making Differentiation a Habit: How to Ensure Success in Academically Diverse Classrooms*. Minneapolis: Free Spirit Publishing.

Kafai, Y. B., Peppler, K. A., & Chapman, R. N. (2010). *The Computer Clubhouse: Constructionism and Creativity in Youth Communities (Technology, Education—Connections Series)*. New York: Teachers College Press.

Middendorf, C. (2009*). Scholastic Differentiated Instruction Plan Book*. Scranton, PA: Scholastic Teaching Resources.

Narvaez, L., & Brimijoin, K. (2010). *Differentiation at Work, K–5: Principles, Lessons, and Strategies*. Thousand Oaks, CA: Corwin Press.

O'Meara, J. (2010). *Beyond Differentiated Instruction*. Thousand Oaks, CA: Corwin Press.

Strickland, C. A. (2009). *Exploring Differentiated Instruction*. Alexandria, VA: Association for Supervision and Curriculum Development (ASCD).

CHAPTER 2

Classroom Assessments

Classroom assessment has a lot to do with gathering information to inform learning and instruction. Teachers make observations and collect information about students in a way that leads to decisions about what to teach and how to teach it. It could be as simple as informed teacher observation or as complicated as using a student portfolio. Sometimes the goal is to guide students toward a specified goal—and sometimes, the goal is to maintain the inventive reasoning process itself.

We will pay special attention to formative assessments—how instructors can assess ongoing learner progress so that subsequent teaching can meet identified learning needs. Such *in*formative assessments can provide immediate and continuous insights into how a lesson is going for individuals and small groups of students. The basic idea is that it's wise to make adjustments *during* a lesson or unit of study.

We will also examine how differentiated learning changes assessment practices and provides multiple entry points to knowledge about math, science, and technology. Other points of interest:

- How assessment-related principles might guide differentiated learning.
- The intersection of activities, assessment, and multiple intelligence theory.
- Lesson planning, adaptive teaching strategies, and formative assessment.
- The exploration of how teachers and students can work collaboratively.

- Suggestions for informative evaluation, performance assessment, and portfolios.

In today's classrooms, technologically savvy teachers often use digital technology to weave *in*formative assessment into the fabric of ongoing instruction. The first step is figuring out how technology will improve assessment and amplify student learning. The goal is to go beyond measuring student progress to adapting new strategies that can enhance learning during the lesson or unit. Possibilities include online assessment software, building a computer-based portfolio, tests using virtual reality, and web-based suggestions from other teachers.

Whether they choose to use technology or not, teachers need to know the difference between where students are and where they need to be. This means consistently observing or checking to see how students are doing as they go through a lesson. Informative assessment requires more than evaluating student progress; an equally important goal is giving students new strategies for being academically successful. As learning tasks are observed and altered, differentiation is a necessity, not an option.

As far as innovation itself is concerned, good new ideas often come from the collision of contradictory thoughts, from conflict, and from trying to do better than what was done in the past. When it comes to preparing students and teachers for the future, it is important to recognize that history is not necessarily cyclical; it is often slow moving and unsteady. Sometimes, progress toward the horizon may seem stationary; at other times, there is sudden acceleration.

ASSESSMENT IS MUCH MORE THAN TESTING

The focus here is on ideas and practical approaches that can help the classroom teacher use assessment to improve instruction. Still, it is important to note that spending too much time on mandated testing has left a sour taste in the mouths of many educators whenever the topic of assessment comes up. This is especially true when it comes to "accountability" and test-related standards. There are, after all, times when standardized testing is more of an end in itself than it is a way to measure or improve student learning.

There have been times when cramming for basic skill tests has gotten in the way of *in*formative assessment and improving the quality of instruction. Still, formative assessments can support standards and are often in line with state and national benchmarks. Whatever side you take in the debate, there is general agreement that it makes sense to act on good information and make quick adjustments during a lesson.

Experienced teachers know that there have always been students who do poorly on traditional tests but show other evidence of learning. Some poor test takers are better at solving problems, getting involved in class discussions, contributing new ideas, drawing sketches, or role-playing what they want to communicate. So the best approach is to provide multiple ways for learners to show their understandings of what they know.

Authentic student assessments, from teacher observations to portfolios, are more engaging and appeal to a wider range of learning styles. When you differentiate assessment, such assessments are usually not about grades. Whether it is before, during, or after a lesson, the basic idea is to figure out what individual students know, understand, and can do. Since formative assessment takes place during a lesson, it makes it easier for the teacher to adjust an ongoing lesson, make it much more interesting, and achieve better results.

With informative classroom assessment, the basic objective is to guide students toward success. Teachers quickly discover that giving students feedback about various stages of their work is more productive than merely giving them grades or normed exams. In the long run, this is more respectful of students and their imaginative possibilities. Clearly, focusing on students' accomplishments can play a crucial role in instructional success.

Assessment can occur at any time—it can be formal, informal, or even accidental. Talking to students in purposeful ways (as you observe their actions) is an informal way of getting useful information. In addition, it is sometimes helpful to take notes about what students are doing as you move around the room. Virtually all students have something to offer as observations are made regarding individual and group strengths and accomplishments.

DIFFERENTIATED INSTRUCTIONAL ASSESSMENT TOOLS

Differentiated instructional assessment is often quite informal. When you see how students work collaboratively with others or alone, you can get a sense of their disengagement or misunderstanding. If learners collect, select, and reflect on their work samples, these can be put into portfolios. Writing in journals is another way to inform teaching and learning.

Differentiated instructional assessment has an important role to play in curriculum planning. When teachers begin to understand the importance of various kinds of assessment, it helps them understand what matters most for students. This is all part of being able to understand what learners need and what they are able to do. When learning goals get adjusted (during a lesson), they can more closely match students' differing needs and enhance academic achievement.

Differentiated instructional assessment is an ongoing process. Instead of just administering an assessment at the end of a unit, good informative assessment also occurs during the lesson. Students who are not grasping the content may need further explanations or demonstrations offered in a different way. Teachers need to know when to reteach and when to move ahead.

As students become better learners, they get a better understanding of learning objectives, the nature of success, and how each assignment contributes to their success. As a sense of self-efficacy connects to feelings of internal control, everyone works harder. And when assessment is viewed as a natural partner in student learning, it becomes a catalyst for better learning and better teaching.

To reach every student requires careful and differentiated planning. As teachers work to engage students with differentiated tools, it can inform the organization of instruction. Success with differentiated tasks requires rules and routines that are carefully selected to meet learners' changing needs.

Some of the differentiated assessment tools teachers can use include

- Providing a comfortable and stimulating learning environment.
- Evaluating students' individual needs before, during, and after instruction.

- Using evaluation data to carefully plan interesting lessons.
- Selecting and organizing activities for the class, individuals, partnerships, and small groups.
- Encouraging each student to continue to learn and improve.

TEACHERS' DECISIONS AND OPINIONS ARE CRUCIAL

A teacher's view when evaluating students' data is very important. When student data is analyzed the teacher makes judgments about how to use the results. With good information in hand, the selection, organization, and pacing of instruction can be determined by students' knowledge bases, backgrounds, strengths, needs, interests, and learning styles.

All teachers have an individual management style that is mirrored in their daily teaching. Knowing their students is a critical part of both their daily routine and how they go about differentiating instruction. Even if you are new to the process, there is no need to throw out planning methods from past years. No matter how experienced you are, it helps to analyze how well you're doing at providing variety and challenge in learning. It also helps to recognize which students are best served by your current plans and altering those plans as needed so all students can be successful.

DIFFERENTIATED LEARNING GOALS

Our knowledge continues to expand as we gain more information about how the brain learns. Our views about effective teaching and learning are based on the latest brain research that informs our work. Following are some of the goals of effective differentiated learning based on brain research. Teachers can

- Involve students in activities that respond to their individual needs, strengths, and inclinations.
- Create challenging and innovative assignments for every learner.
- Present management strategies to include flexible grouping approaches to content and instruction.

- Focus on students' readiness, instructional needs, and learning preferences.
- Aid in the process of gathering and organizing assessment data.

It always works out better if teachers choose and adapt a wide range of ideas and suggestions as they go about working at the intersection of assessment and instruction.

MULTIPLE PATHS TO LEARNING AND ASSESSMENT

Teachers have always been faced with the challenge of dealing with the fact that individual students learn things in different ways. Yet tracking didn't seem to do anybody much good. Differentiated learning is a proven way to assist with mixed-ability group instruction—while at the same time, recognizing that students learn in unique ways and at varying levels of difficulty.

Recognizing individual learning styles and adapting a differentiated teaching style can make learning easier. With differentiated learning, the teacher provides specific ways for each student to learn deeply, working energetically to ensure that all students work harder than they thought possible and achieve more than they imagined.

WHAT TO LOOK FOR IN PLANNING LESSONS

Readiness: Student readiness is their current knowledge, understanding, and skill level. Supplemental books can be provided at different reading levels. Small mixed groups of students can do collaborative activities at various levels of difficulty that are focused on the same learning goal.

Interest is another way to differentiate instruction. Teachers often let students choose from a variety of media arrangements such as video, music, film, and computers to express their ideas. They invite students to write about their favorite movies and encourage them to examine their thoughts about the music that excites them. Computer models also help support the students' topic of interest. Teachers can vary the activities based on the interests and the aptitudes of their students. Everyone reaches the same concept although he or she may go about it differently.

A student's *learning profile* or *learning style* draws on his or her academic and cultural background. Teachers try to develop activities that use many viewpoints on interesting topics. Have students present a project in a visual, auditory, or movement style. Being able to base at least some math, science, and technology instruction on a student's preferred way of learning has proven to be especially helpful in teaching a wider range of students.

According to the national educational standards, "The study of math, science and technology should include many opportunities for communication so that students can

- reflect on and clarify their thinking about math, science, and technology ideas and situations;
- realize that representing, discussing, reading, writing, and listening to math, science, and technology are an important part of innovative learning;
- model situations using oral, written, and active learning methods;
- evaluate math, science, and technology ideas" (National Council of Teachers of Mathematics, 2000; National Academy Press, 1996; ISTE, 2008).

A MULTIPLE INTELLIGENCE MODEL

Lesson ideas for selected intelligences are illustrated below.

Linguistic Lesson Ideas

1. Have students find a pattern in their math, science, or technology explorations.
2. Students communicate the pattern by writing, in two complete sentences, what they observed. Describe the pattern in general terms.
3. Students describe, step by step, the procedures they used to arrive at a solution.

Spatial Lesson Ideas

1. Have students make a diagram that shows the math, science, or technology procedures.

2. Students construct a flow chart that shows the steps they must do to complete the task (for example, measure the distance).
3. Learners use manipulatives to model math, science, and technology concepts and skills.
4. Encourage the class to draw a picture or make a design of something they are trying to remember.

Musical-Rhythmic Lesson Ideas

1. Play three to five minutes of classical music before starting learning tasks.
2. Use sound breaks between activities to energize tired energy and low moods.
3. Use clapping. singing, humming, rhythmic movement, rhythmic words (raps), or jingles to help students recall and recite concepts. Example: sing the fours table to "Jingle Bells" (starting with "Dashing through the snow . . . 1 times 4 is 4, [and a] 2 times 4 is 8, 3 times 4 is 12, four 4s are 16, five 4s are 20").

Kinesthetic Lesson Ideas

1. Linking cubes and base ten blocks can be used to model math concepts of place value and regrouping.
2. Pattern blocks can also be used to create spatial geometric learning.
3. Science rock searches or leaf explorations use this intelligence. Students describe and gather natural objects.
4. Play physical games to practice math, science, and technology concepts. Hopscotch or Twister designed for math facts practice works well. Card games and board games like Concentration or Scrabble redesigned for use with number cards and markers also works. Students can help with the design and rules of the game.

Interpersonal Lesson Ideas

1. Use collaborative learning techniques often. They will enhance students' learning experiences.

2. Students can work together to create five-minute commercials about outer space, the moon, black holes, or faraway planets.
3. Encourage students to play two- or three-person math and science games to practice basic skills (card games such as forehead poker or Go Fish can be used).

Logical-Mathematical Lesson Ideas

1. Encourage mental math daily. Whenever possible, request that students attempt to do the simpler calculations in their heads.
2. Teach estimation skills. "How much do you expect the answer to be?" For example: 24×18 (think: $25 \times 20 = 500$). Five hundred would serve as a good guess.
3. Use natural objects to model math, science, and technology concepts and processes. (Examples: use petals, small rocks, or hard berries as counters to model sorting, counting, subtraction, place value, averaging, and so on.)
4. Use outdoor settings to teach math and science concepts. For example: "Estimate the number of needles on this evergreen tree. Back your estimates with counts and calculations that justify your guesses."
5. Use sorting and classifying as a way to make math, science, and technology terms understandable. Within this framework, logical-mathematical intelligence might be defined as the ability to solve problems, generate new problems, and do things that are valued within one's own culture.

Multiple intelligence (MI) theory suggests that these eight "intelligences" work together in complex ways. Most people can develop an adequate level of competency in all of them. And there are many ways to be "intelligent" within each category.

COGNITIVE RESEARCH POINTS TO MULTIPLE ENTRY POINTS TO KNOWLEDGE

Learning has a lot to do with finding your own gifts. Many questions remain, but no one doubts that today's students are a complex lot, with varying needs, abilities, and interests. To make learning more accessible to such a wide range of students means respecting multiple ways of making meaning. The brain has a multiplicity of functions and voices that speak independently and distinctly for different individuals (Tomlinson, 2010).

Obviously, no two children are alike. An enriched environment for one is not necessarily enriched for another. The basic idea is to maximize each student's learning capacity. We use the term *differentiated instruction* to refer to a systematic approach to teaching academically diverse learners. It is a way of thinking about students' learning needs and enhancing each student's learning capacity.

Good informative assessment techniques can help teachers become aware of who their students are and how student differences relate to what is being taught. The insights provided by authentic assessment techniques can help teachers adapt student assignments and increase the possibility that each student will learn.

The unique aspect of the brain is that its cells are arranged in such a way as to make the brain malleable or plastic enough for learning to occur and stable enough for the learning to solidify into wisdom. Since a steady stream of new brain cells is continually arriving to be integrated into new circuitry, then the brain is more malleable than anyone had realized. "Our brains remain remarkably plastic and we retain the ability to learn throughout our lives" (Jensen, 2005).

Neuroscientists point out that the brain is extremely plastic and dynamic, very responsive to experience. Various studies also suggest that a small, well-connected region of the brain is in charge of organizing and coordinating information, acting like a global workspace for solving problems. Each year new research on brain functioning adds to our understanding of how the mind functions. Although new cognitive insights help us understand our students, we still need more assistance in directly applying these findings to classroom practice.

Assessment can be used to motivate students in several ways. To begin with, students can form collaborative teams for interdisciplinary inquiry and peer assessment. As students are brought into designing assessment procedures as responsible partners, the whole process is enhanced. Students can then use portfolios to keep their own records and reflect on how well they are doing.

By viewing the evidence of their increasing proficiency, they reflect on their own progress. By learning to communicate with peers, teachers, and parents about their achievement, learners can also learn to take more responsibility for their own academic success (Costantino, De Lorenzo, & Kobrinski, 2006).

PORTFOLIOS: THE INTIMATE ASSESSMENT TOOL

Portfolios are formative performance assessment tools for having students assess how well they are doing. Formative performance assessment portfolios can be used by teachers to modify or adapt instruction based on student performance. Students produce evidence of their progress to adjust their own learning.

Performance portfolios help teachers create performance tasks tailored to the maturity of their students. These portfolios provide a framework for teachers to have conversations about the strengths and weaknesses of student performance. Performance portfolios can also be used to describe the level of performance, referred to as a "rubric" (Expert, Proficient, Novice, Beginning) of an individual student. For example, the goal may be for all students to reach the Proficient level at the end of fifth grade. Portfolio assessment provides individual attention to students who are making insufficient progress toward that goal.

Benefits of Using Portfolios

Portfolio assessment in science and math has the following advantages:

- Assessment helps make connections and communicates needs.
- Portfolios organize information and provide authentic information.
- Portfolios are an effective way to communicate with students, parents, and administrators.
- Portfolios emphasize a student's strengths and interests.
- Portfolios encourage students to assume responsibility for their learning.
- Portfolios measure student achievement over time.

If the purpose is to assess daily, ongoing work, students will actively participate in planning, selecting criteria, and evaluating their portfolios. These possibilities create feelings of pride and ownership by the students.

Portfolios provide a window on student learning and classroom teaching that can't be generated by other assessment tools. They have also proven themselves useful for viewing a student's interactive skills, thinking, and competency in science and math. When teachers share the purpose of portfolio construction with their classes, the students are more likely to see the value and relevance of selecting and reflecting on their important work.

Sharing their portfolio work with others is very motivational and everyone gains. For teachers, sharing might involve collecting and organizing data, graphing the results, presenting a teaching strategy, and having an informal discussion of authentic assessment with a colleague. Remember, when you set out to improve something, it's important to have a goal in mind. And it is just as important to leave room for serendipity. When it comes to possibilities for improving practice, we have to be ready for chance finds.

STUDENT SELF-ASSESSMENT USING RUBRICS

An important element of formative assessment is feedback. Finding the time to provide feedback to every student is difficult for many teachers. Often a forgotten component, students are an excellent source of feedback. Student self-assessment can provide a useful and accurate source of feedback. Students are often the best sources to reflect on the quality of their work, judge how it meets the goals or criteria, and revise if necessary.

Self-assessment is formative (or "informative" as Carol Tomlinson [2008] describes it). Students access their work in progress to find ways to improve their performance. Provide a rubric or create one with your students. A rubric lists criteria and describes the varying levels of quality from excellent to poor. When carefully designed with collaboration by students, good rubrics can provide students with clear guidelines without limiting creativity.

Teacher-Made Rubric

1. *Set clear expectations.* Expectations should be clearly defined by the teacher, the student, or both. List the qualities of each category. Describe several skills for each section. Descriptors should be specific. Decide on the number of points each skill is worth. This should model the state and national standards.
2. *Conduct thoughtful self-assessment.* Test the rubric with your students and check to see if it flows and makes sense.
3. *Revising is a crucial step.* Students need to know their efforts can lead to actual improvement opportunities.
4. *Student attitudes change with self-assessment.* They tend to value it. Students reported they could self-assess effectively when they knew what the teacher expected and made attempts at revising their work.

Self-Assessment

Self-assessment can be used in any subject. If students are involved in its creation, they can assess and improve it. A way to begin is by using the following self-assessment form:

Student Self-Assessment Form

Name:
Date:

Please answer the following questions in a thoughtful and truthful way.

1. What math, science, and technology skills are you good at now?
2. Which math, science, or technology skills would you like to be better at?
3. What could you do to improve your work in math, science, or technology?
4. Have you found more than one way to solve a problem?
5. What do you like about your work in math, science, or technology?
6. What have you learned about working with others?

After students complete the self-assessment form, their answers can be shared with other students, and their work can be edited and placed in their portfolio.

CONNECTING ASSESSMENT WITH INSTRUCTION

Portfolios are proving useful in connecting assessment with instruction at every level because they allow students and teachers to reflect on student learning. They pay attention to students' ideas and thinking processes while offering opportunities for looking specifically at how students are doing. With the emphasis on portfolios, we're not suggesting that more traditional testing has no place in the curriculum. Rather, we must respect the limits of traditional testing methods and search for more connected measures of intellectual growth.

When added to other performance measures, like projects, writing assignments, presentations, and field trips, portfolios make an important contribution to differentiated instruction.

GROUPING IN THE CLASSROOM

Managing grouping strategies is part of your teaching role. An important management task in the differentiated classroom involves making grouping decisions. Students can be brought together as a class when they need to receive directions, discuss ideas, learn new skills, or practice before a test. Students may work alone or in small groups based on their current needs. Planning for grouping requires analyzing preassessment results, arranging the classroom, and allotting time.

Students need specific directions whether dividing the class into small groups or allowing them to do independent work. When learners are able to carry out activities with little or no adult supervision, the teacher can assess and observe students or provide direct instruction to individuals.

Flexible Grouping

Flexible grouping allows students to learn information with the whole class, individually, with a partner, or in a small group. The teacher selects the grouping strategy and decides the grouping choice.

Flexible groups try to meet the identified academic, social, and emotional needs of each student. Groups can be formed by students with similar interests, which enables them to combine their experiences and share their excitement. When each student engages in a variety of grouping decisions, he or she learns to work independently and collaboratively with others. Teachers plan instruction by identifying students' needs and assess each individual using instructional (formative) assessment.

FORMATIVE ASSESSMENT SUGGESTIONS

Formative assessment is the assessment of student learning integrated into the act of teaching. Formative instructional assessment takes place during instruction. It provides feedback to teachers and students and allows them to make adjustments that will improve their achievement and engage learners in a variety of flexible groups. It is important to avoid old grouping ruts by mixing up the grouping designs, making student-choice and teacher-assigned group arrangements available.

All students need to know what is expected of them in the classroom. When working in collaborative teams, students should be encouraged to try out some formative assessment strategies. For example, when studying a unit on plants with second graders, have the team members divide up the parts of a plant according to their characteristics (stem, flower, leaves, roots, etc.). Next, have them describe each part and explain how the plant uses it.

Students can assess their learning by illustrating the plant and explaining it to a partner. This is an ongoing formative assessment. Students collect information before and during the class; then they write about their observations, guesses, and thoughts about their experiences. This makes the lesson very hands-on and authentic for both the students and the teacher.

Older students also use formative assessment. A class of fifth graders can follow similar procedures when studying rocks. Students try to find out which of the rocks is sedimentary, igneous, or metamorphic. They read and follow the information found in their textbook, examine their rocks, discuss rock characteristics with their team, plan, compare, test the rocks, describe each rock, illustrate the rocks, and explain how they decided which rock characteristics belonged to each rock. Writing and discussing are important parts of formative assessment.

DIFFERENTIATED ASSESSMENT

In a differentiated math/science and technology classroom, teachers use whatever combination of tools that work to teach and assess the progress of all students. Examples include experiments, collaborative inquiry, graphic organizers, problem-solving worksheets, calculators, and concrete materials that children can manipulate to try out different numeric relationships. From simple observation to personal portfolios, differentiated performance assessment can be used to motivate students.

Questions for the teacher to consider:

- What questions will most stimulate students' thinking?
- Can the students understand, organize, and systematically use data?
- Do students comprehend the facts, concepts, and solution to problems?
- What are the differences in the ways students present their conclusions?
- How can students improve their communication in math, science, and technology?

All teachers want students to be motivated and responsible; most believe that students should ask questions and struggle with problems to find reasonable answers. Many teachers also share a vision of what they think should be happening in their classes. This includes tailoring instruction and assessment to meet the needs of various students and clusters of students. It also includes having students work together in small groups so that they can undertake investigations and accomplish tasks. Examples of related learning tools include manipulative materials, blocks, beakers, clay, rulers, chemicals, musical instruments, calculators, assorted textbooks, computers, the Internet, and more.

Once teachers have some idea of a student's strengths, they can differentiate lessons in a way that builds on that student's strengths. Sometimes, the next step is making sure that a student works with a partner who has a different profile. At other times, common interests might influence who comes together in a group of three or four.

ENGAGE, EXPLORE, EXPLAIN, ELABORATE, AND ASSESS: STRATEGIES FOR INNOVATIVE LEARNING

- When students are *engaged*, they might focus on things like current events, local issues, scientific demonstrations, experiences on a field trip, or a question that students will encounter in the unit. Connections are made between past and present learning experiences.
- When students *explore,* they may make observations, collect data on the Internet, or carry out investigations using laboratory equipment. Students collaboratively explain their ideas to others. They build a common base of experiences and actively explore their environment.
- When students *explain*, they develop different rationales and use scientific terms. The content can make earlier experiences easier to describe and explain.
- When students *elaborate*, they extend the information in new contexts. Opportunities are given for deeper and broader understandings.
- When students *assess* they are encouraged to look at their understandings; they apply what they learned. They can place evidence of how they apply the information in their math/science and technology portfolios.

Rules for Assessing

1. Conduct informal student-centered assessment in the context of learning teams. You must assess each student's achievement, but it's far more effective when it takes place in a collaborative setting. When it comes to grading judgments, it is best to avoid group grades.
2. Give continual feedback and assessment. Learning groups need continual informal feedback on how each member is doing. Everything from simple observation and presentations, to answering questions and oral presentations needs to be considered.
3. Develop a list of expected behaviors:

Prior to the lesson _____

During the lesson _____
Following the lesson _____

4. Directly involve students in assessing their own learning and the group work. Group members can provide immediate help to maximize all group members' learning.
5. Avoid all comparisons between students that are based solely on their academic ability. Such comparisons will decrease student motivation and learning.
6. Use a wide variety of lesson plans and assessment tools so that lessons become a vehicle for growing skill and knowledge acquisition.

ASSESSMENT AND THE SOCIAL NATURE OF LEARNING

When teachers differentiate and arrange for their students to spend significant amounts of time in learning groups, their assessment techniques change dramatically. Within this context, more emphasis is placed on student understanding and less on content recall assessed through multiple choice tests. When you plan for both differentiated instruction and assessment at the same time, you improve both.

Assessment becomes more authentic and dynamic as it changes to reflect the changing nature of instruction in the classroom. Portfolios, for example, document performance and provide tangible evidence of learning. So it should come as no surprise to find that the content of students' portfolios can be used to guide differentiated instruction as well.

Since effective performance assessment is an essential ingredient in the differentiated classroom, suggestions in this chapter are designed to be part of the differentiated teaching and learning process. As an interactive learning tool, performance assessment, including portfolios, can help teachers assess their students continually, both informally and formally.

In the context of math, science, and technology instruction, performance assessment is viewed as a well thought out opportunity to demonstrate competency. Developing mathematical power, using science processes, and doing innovative research involve more than doing scientific inquiry. It calls for giving students increasingly difficult math, science, and technology problems to solve.

The social context of students' performance is an important issue in teaching challenging activities. The amount of help that children need can be adjusted by observing and analyzing the interaction that occurs during the learning task. As students gain social and subject matter competence, the teacher can provide less assistance and shift more responsibility to the student groups.

The following are suggestions for getting started with differentiated instruction and related performance assessment practice. At the beginning of each unit of science or math study, teachers can measure students' understanding and establish learning objectives by writing them on a chart entitled "What our class wants to find out." Next, teachers could introduce a variety of hands-on activities that have a range of differing possibilities for every small cluster (group) of students.

Although the goal is to have all of the students reach for the same standard, it is still possible for them to go about it differently. If you try to get students to learn the same things at the same time and in the same way, too many will fail to reach the standards set for a lesson. Portfolios are a natural partner for DI because everyone collects, selects, and reflects information on the same set of skills and competencies. It's just that no two portfolios are ever done exactly the same.

INFORMED TEACHERS HOLD THE KEYS TO THE FUTURE

We can't predict the future based on the past, but there is a lot to learn from past experiences. Traditionally, there has been a lack of coherence among system, school, and classroom approaches to assessment. The educational past also suggests that top-down directives and standardized tests have had little influence on the quality of student learning. As a consequence, we have paid more attention to things like observation, portfolios, and formative assessment in the differentiated classroom.

High-visibility summative tests may be open to question, but good informative assessment is crucial to achieving the goals of instructional practice. Differentiated instruction and related assessments are built on the teacher's ability to figure out a student's prior experiences, multiple intelligences, developmental needs, personal preferences, and levels of subject matter competency.

At their best, portfolios and formative and related performance assessments require students to demonstrate the desired procedure or skill in a realistic context. Such techniques allow students—or teachers for that matter—to display growing strengths rather than simply exposing weaknesses. By taking an approach that connects DI to authentic assessment, teachers promote more student ownership of learning. And by encouraging students to use, shape, and reflect on the science and math knowledge that is most important, they can amplify both assessment and instruction.

Generating the energy to act on new practices is an art. It has a lot to do with understanding human nature, establishing a readiness for change, and developing an intellectual understanding of new possibilities. The most effective innovations are usually research based and classroom friendly. Change is a personal process. So making changes in curriculum and assessment without involving teachers is a contradiction in terms.

Moving assessment and quality math/science/technology instruction from talk to action requires the involvement of informed teachers who can make good use of new concepts. In addition, teachers must be involved in educational decision making. If their voices are left out, then our schools will miss the important improvement opportunities provided by innovations like differentiated instruction and portfolio assessment.

Professionals don't blossom when they spend their careers enfolded in the logic of others. Teachers themselves are in the best position to identify the key issues, question, probe, assess, and try to attain clarity. They are also the ones who can implement differentiated learning in the sometimes ambiguous world of science and mathematics instruction. When it comes to pedagogical change, informal belief systems are just as important as methodology.

As teachers become students of their own learning, they can discover the inconsistencies between what they believe about gaining knowledge and how they practice teaching. After all, the discoveries we make for ourselves are more convincing and make us more willing to change what we do. In many ways, the thoughtful reflection about practice that comes with using good assessment tools and differentiated instruction helps teachers become more effective professionals.

Even while looking for factors that get in the way of student achievement in the future, it is possible for teachers to create classroom assessment routines that support high-quality instruction today. When it comes to understanding underachieving students, we need to talk more about responsibility and less about blame. It will take serious discussions among serious people seeking serious solutions to move reluctant learners in the direction of viewing education—and the hard work associated with it—to light up students' future.

SUMMARY AND CONCLUSION

In just about every classroom, there are students who have trouble learning some subjects. For many, mathematics and science are at the top of the aggravation list. And yes, there are students who have trouble with digital technology. But whatever the subject or topic, some students are more motivated than others. To stay on the positive side of the equation, everyone in the classroom has a responsibility to value inventive, self-reliant, and motivated behavior. Making this happen can be an exciting adventure for everybody.

Differentiated instruction is a good way to meet specific individual and small group needs in the regular classroom. It is an organized, yet flexible, way of adjusting teaching and learning to meet students where they are and help them achieve. In a differentiated classroom, teachers often use instructional strategies that build on multiple intelligence theory, cooperative group work, and portfolio assessment to meet unique learning and assessment needs.

A wide range of assessments are needed so that teachers and students have good information about where learners are and where to go next in the learning process. Formative assessments are particularly useful in this regard, especially when used with differentiated teaching strategies. A reminder: formative assessment provides information about how learners are progressing and how well they understand what they are working on. Besides tailoring what is being taught to meet identified needs, the teacher has to balance instructional and assessment goals to meet a variety of individual and group needs.

Understanding the scientific method and related experimental processes opens an infrequently used door to good assessment and instruction. The fusion of these elements provides opportunities for cultivating self-awareness. And as far as generating new ideas is concerned, scientific reasoning can also support going beyond received opinion and gaining an understanding of the reality behind natural appearances.

When authentic assessment is coupled with differentiated learning, teachers become more effective at identifying the needs of all their students. This makes it more likely that ongoing lessons can be adapted in a way that energizes each student. The feedback that comes from various forms of informative assessment helps teachers reflect on their work and come up with constructive and energetic approaches for maximizing student learning.

Programmatic elements like the curriculum, classroom assessment, and class size are all important. But the quality of teaching matters even more. Teachers are, after all, the ones who must establish an environment in which students constantly learn, grow, and become better at what they do. Understanding the characteristics of effective instruction certainly helps. But, unfortunately, no one seems to be able to clearly define what good teaching is. So *developing the skills of all teachers* is the key to replicating the successful assessment routines that are part of high-quality instructional programs.

The overall assessment policy goal: to make sure that teachers can effectively use the latest assessment methods to inform instruction. As far as the classroom is concerned, success has a lot to do with developing a good understanding of the point of view of others. So doing a good job of integrating assessment and instruction depends on being able to go beyond your point of view to figure out what others think.

The objective for teachers: mastering the art of teaching in a way that lights subjects like math, science, and technology by the bright beams of their students' imagination.

REFERENCES

Costantino, P., De Lorenzo, M., & Kobrinski, E. (2006). *Developing a Professional Teaching Portfolio: A Guide for Success*, 2nd ed. Boston: Allyn & Bacon (Pearson Education).

Jensen, E. (2005). *Teaching with the Brain in Mind*, 2nd ed. Alexandria, VA: Association for Supervision and Curriculum Development.

National Academy Press. (1996). *National Science Education Standards.* Washington, DC: National Academy Press.

National Council of Teachers of Mathematics (NCTM). (2000). *Principles and Standards for School Mathematics.* Reston, VA: NCTM.

Tomlinson, C. A. (2008). Learning to love assessment. *Educational Leadership,* 65(4): 8–13.

RESOURCES

Abel, S., & Volkmann, M. (2006). *Seamless Assessment in Science: A Guide for Elementary and Middle School Teachers.* Portsmouth, NH: Heinemann.

Aeration, P. W. (2000). *Classroom Assessment: Concepts and Applications.* New York: McGraw-Hill.

Bransford, A., Brown, L., & Cocking, R. (Eds.). (1999). *How People Learn: Brain, Mind Experience.* Washington, DC: National Academy Press.

Brookhart, S. M. (2010). *Formative Assessment Strategies for Every Classroom: An ASCD Action Tool.* Alexandria, VA: Association for Supervision and Curriculum Development.

Burch, G., & Timber, M. (2004). *Educational Assessment for Elementary and Middle School Education.* Upper Saddle River, NJ: Pearson Education.

Crassly, J. (2006). *Developing Assessment Literacy: A Guide for Elementary and Middle School Teachers.* Norwood, MA: Christopher-Gordon Publishers.

Gardner, H. (1983). *Frames of Mind.* New York: Basic Books.

———. (1993). *Creating Minds.* New York: Basic Books.

———. (1997). Multiple intelligences as a partner in school improvement. *Educational Leadership* 55(1): 20–21.

Greenstein, L. (2010). *What Teachers Really Need to Know about Formative Assessment.* Alexandria, VA: Association for Supervision and Curriculum Development.

Gregory, G., & Kuzmich, L. (2004). *Data Driven Differentiation in the Standards-Based Classroom.* Thousand Oaks, CA: Corwin Press.

Hales, L., & Marshall, J. (2004). *Developing Effective Assessments to Improve Teaching and Learning.* Norwood, MA: Christopher-Gordon Publishers.

Hamm, M., & Adams, D. (2009). *Activating Assessment for All Students: Innovative Activities, Lesson Plans, and Informative Assessment.* Lanham, MD: Rowman & Littlefield Education.

Heacox, D. (2002). *Differentiating Instruction in the Regular Classroom: How to Reach and Teach All Learners, Grades 3–12.* Minneapolis: Free Spirit Publishing.

Heritage, H. M. (2010). *Formative Assessment: Making It Happen in the Classroom.* Thousand Oaks, CA: Corwin Press.

ISTE (International Society for Technology in Education). (2008). The ISTE NETS and Performance Indicators for Teachers. Retrieved November 30, 2010, from http://www.iste.org/standards/nets-for-teachers.aspx.

MacMillan, J. H. H. (2000). *Classroom Assessment: Principles and Practice for Effective Teaching.* Boston: Allyn & Bacon.

Montgomery, K. (2000). *Authentic Assessment: A Guide for Elementary Teachers.* Boston: Addison-Wesley.

Nikko, A. J. (2000). *Educational Assessment of Students.* Upper Saddle River, NJ: Prentice Hall.

Popham, W. J. (2010). *Everything School Leaders Need to Know about Assessment.* Thousand Oaks, CA: Corwin Press.

Salvia, J. (2001). *Assessment.* Boston: Houghton Mifflin.

Shute, V. J., & Becker, B. J. (2010). *Innovative Assessment for the 21st Century: Supporting Educational Needs.* New York: Springer.

Sternberg, R. (1988). *The Triarchic Mind: A New Theory of Human Intelligence.* New York: Viking Press.

Stiggins, R. J. (2000). *Student-Involved Classroom Assessment.* Upper Saddle River, NJ: Prentice Hall.

Tobey, C. S., & Minton, L. (2010). *Uncovering Student Thinking in Mathematics, Grades K–5: 25 Formative Assessment Probes for the Elementary Classroom.* Thousand Oaks, CA: Corwin Press.

Tomlinson, C. A., Brimijoin, K., & Narvaez, L. (2008). Differentiated School: Making Revolutionary Changes in Teaching and Learning. Alexandria, VA: ASCD.

Tomlinson, C. A. & Imbeau, M. B. (2010). Leading and Managing a Differentiated Classroom. Alexandria, VA: ASCD.

Tomlinson, C., & McTighe, J. (2006). *Integrating Differentiated Instruction & Understanding by Design: Connecting Content and Kids.* Alexandria, VA: Association for Supervision and Curriculum Development.

Warmly, R. (2006). *Fair Isn't Always Equal: Assessing and Grading in the Differentiated Classroom.* Portland, ME: Stenhouse Publishers.

Assessment Training Institute (www.assessmentinst.com) and www .makingstandardswork.com are two websites we often use to explore assessment issues and practices.

CHAPTER 3

Learning Mathematics
Problem Solving, Creativity, and Innovation

This chapter highlights *problem solving* as key to helping students make sense of mathematics. Mathematical reasoning skills are viewed as an integral part of the process. We also point to ways that imaginative thinking, differentiation, and collaboration can serve as useful platforms for building mathematical knowledge. Major ideas from the math standards are emphasized.

Principles and Standards for School Mathematics (NCTM, 2000) is a resource and guide for everyone involved in making decisions that affect mathematics education. It stresses the importance of learning how to use reasoning strategies to solve problems that arise in mathematics and other contexts. The standards document also suggests that innovative tendencies can be nurtured during the process of learning mathematics.

Although it has not been well understood, human ingenuity has always been highly valued in America. Increasingly, nations around the world have been making creative development a societal priority. The potential consequences of creativity and innovation seem clear to just about everyone. But arriving at original and useful ideas is easier to talk about than it is to actually do.

Creativity requires constantly moving between generating many unique ideas (divergent thinking) and combining those ideas in a way that provides something useful (convergent thinking). Idea finding, solution finding, teamwork, and a plan of action all have a role to play in moving between divergent and convergent thinking. Also, fact finding and collaborative research are important checkpoints along the road to creativity and innovation.

Imaginative breakthroughs in the field of mathematics rarely come from an individual working in isolation. Imaginative ideas and new approaches don't happen in a vacuum. Innovative mathematicians build on the work of others, although creativity and teamwork often go hand in hand. And collaboration is enhanced when everyone involved knows enough about what is being studied to combine elements in different ways.

In one way or another, we all have to deal thoughtfully and responsibly with numbers. If we don't, the results can be catastrophic. Economic failures, for example, are often brought on by an uninformed addiction to unnecessary risk. Like gambling, it is for the mathematically challenged. Even when catastrophes like earthquakes and oil spills occur, lessons can be learned when the contributing elements are uncovered, documented, and reworked to make improvements.

As they go about solving problems, teachers *and* students soon learn that there is more than one way to ask questions and come up with answers. Whether it is the master or the apprentice, neither one size nor one approach fits all. Einstein and Thoreau are good examples at the high end of the scale. Both explored the natural world. They followed quite different paths, but they both had a willingness to ask questions, to explore, to study, and to wonder. We can all do at least a little of that.

BECOMING FULLY ENGAGED IN THE MATHEMATICAL PROCESS

Building on a powerful set of ideals makes sense as long as there is room for thoughtful change. When teachers are able to organize their mathematical understandings around key principles and imaginative concepts the results are bound to rub off on their students. Along the path to mathematical knowledge, it doesn't take long to realize that the focus should be on logic and reasoning, rather than authority and convention.

Print literacy and mathematical literacy (numeracy) have a lot in common. Like reading and writing, everyone has to reach some level of competency, but only a few will be award-winning novelists, journalists, or scientists. And even though most students are not going on to be mathematicians, they can all emulate their intellectual curiosity and reasoning skills.

Young learners *can* ask questions and solve math-related problems. Under the right conditions, it is also possible for them to become enthusiastic as they go about gaining a better understanding of mathematical implications of the world around them. Pressure to become compliant learners who spend most of their time answering teacher-formulated questions is a slippery slope in the direction of boredom and unimaginative responses.

For many educators, questioning and inquiry are the ideal paths to mathematical problem solving, igniting student interest, and generating inventive ideas. But the will to do innovative work cannot be imposed by others. To be self-starters, motivated, and more successful, students need an active environment where full engagement is possible.

Creative, thoughtful learners tend to raise more questions, inquire more deeply, and offer alternate points of view. When exploring an issue, too often learners look for simple answers, whereas innovative students search for additional information. At any age or skill level, monitoring and reflecting on the process of mathematical problem solving is central to generating new ideas.

For students to move beyond someone else's words, ideas, and solutions, they need opportunities to struggle with a task that inspires and motivates them. Most elementary and middle school math teachers today soon discover their classrooms are filled with a diverse group of students who have varying abilities and learning styles. What can teachers do to help all students succeed? Differentiated learning is part of the answer.

Some students have an aversion to mathematics and feel they would be better off if they could avoid it. As a result, all kinds of misguided ideas stem from policy makers and citizens who don't understand when it comes to applying math principles.

Not so many decades ago, math instruction focused primarily on the computational skills of arithmetic along with whole numbers, fractions, decimals, and percentages. Math basic skills still matter, but, today, there is agreement that a deep understanding of mathematics is more than knowing facts, figures, and computation (NCTM, 2010).

The challenge for teachers today is to teach the basics along with mathematical reasoning and problem-solving skills. To do this well requires helping students develop a positive and confident attitude toward the subject.

Typically, students bring varying backgrounds to their math lessons and teachers work hard to accommodate this diversity. We believe that all students have the potential to learn mathematics. We have to admit that at least some of the students arrive so unprepared that they encounter academic difficulties. Everyone from math educators to textbook writers have been working hard to develop creative and innovative ways to meet the mathematical needs of all students.

EXPLORING MATHEMATICS

What do you think mathematics is? Many people have their own ideas. Examine your thinking about mathematics.

1. *Math Is Thinking and Asking Questions.* How you make plans, organize your thoughts, analyze data, and solve problems is *doing* mathematics.

Student Math Activity: List all the ways you used math this week.

2. *Math Is Finding Patterns and Making Connections.* Students need to understand math concepts and make connections with ideas they know. Students soon learn how the basic facts of addition and subtraction are interrelated.

Student Math Activity: First and second graders can show how the math combinations ($4 + 2 = 6$ and $6 - 2 = 4$) compute. In later grades, students demonstrate the connections between multiplication and division. They use their observation skills to describe, classify, compare, and solve problems.

3. *Math Is a Tool.* Students learn that the tool of mathematics is used by everyone every day. Like mathematicians, scientists, and technology experts, students use math tools to solve problems.

Student Math Activity: Solve this problem using the math tool of problem solving. Pretend you bought a horse for $50 and sold it for $60, and then you bought the horse back for $70 and sold it again for $80. How much money did you make or lose? Do the problem with a small group and explain your reasoning.

4. *Math Is Fun (Solving a Puzzle).* Anyone that has ever worked on a puzzle or a stimulating problem knows what we're talking about when we say mathematics is fun. The stimulating quest for an answer encourages you to find a solution. Innovation fits easily into this process.

Student Math Activity: With a partner, play a videogame of your choice. Many card games and videogames are innovative math challenges. Keep track of points.

5. *Math Is a Creative Art.* Many students think of math as a confusing set of facts and skills that must be memorized. They need to be dissuaded from this view and instead taught that math can be appreciated as an art form where everything is related and interconnected.

Student Math Activity: With a group of students, design a picture with geometric shapes, label them, decorate them, and add to them.

6. *Math Is a Specialized Language.* The language of mathematics uses special terms and symbols.

Student Math Activity: Get students into groups of three or four. Have the group members decide on a question or a problem they would like to investigate. Once a question is agreed on, have them organize and gather data. Students will present the data by making a clear, descriptive chart or graph that can be shown to the class.

7. *Math Is Part of Many Subjects.* Mathematics relates to many subjects. Literature, science, music, art, social studies, physical education, movement, and just about all subjects make use of mathematics in some way.

Student Math Activity: With a group, create a song using a rhythmic format. It can be sung, chanted, or rapped. The lyrics can be written and musical notation added.

MATHEMATICAL INQUIRY AND COLLABORATION

Collaborative inquiry focuses on students' interaction and natural curiosity. Lesson activities can involve pairs of students or small groups. Through discussions, students become better critical and creative thinkers. This active process is a powerful force for encouraging students to ask questions, gather data, observe, analyze, propose answers, explain, predict, and communicate their results (Stephen, Bowers, Cobb, & Gravemeijer, 2004).

The teacher's challenge is to organize group work that engages students in meaningful math activities. All students need to be challenged to think and work together to solve problems. The next step is helping them feel secure as they go about debating and imaginatively applying their understandings.

All students need to be involved in quality, engaging mathematics instruction. High expectations are set for all learners, but accommodations have to be made for struggling students. Students confidently engage in mathematics tasks, explore evidence, and provide reasoning and proof to support their work. As active problem solvers, students will be flexible as they work in groups with access to technology. Students value mathematics when they work productively and reflectively as they communicate their ideas orally and in writing.

This is just part of the vision set forth in the National Council of Teachers of Mathematics 2000 standards document. Here we set out to help teachers gain a better understanding of the mathematics standards. Along the way, suggestions are made for blending creative and innovative skills into effective mathematics teaching.

A VIEW OF THE NATIONAL COUNCIL OF TEACHERS OF MATHEMATICS STANDARDS

The NCTM standards are descriptions of the math content and processes that all students should learn. They call for a broader scope of mathematics studies and describe an in-depth overview of what students should know and be able to do. The content standards describe what you are teaching. They do not tell you how to go about teaching the content. *You* design lessons covering the content standards.

Applying the Curriculum Standards

Here we present a few sample activities for each standard. The intent is not to prescribe an activity for a particular grade level; rather, activities are presented that can be used in many grades.

Number and Math Operations Standard

Skills and student ideas related to numbers are a good place to start. Making sense of numbers and seeing how they are used every day is called *having number sense.* There are various levels of number sense.

Level 1

Most young children don't understand quantity. They may be able to count and identify a number. They can point to the numeral 2, but do not know the actual meaning of the number at this level.

Level 2

At this level, a child starts to understand number sense. They talk about words like "many," "a hundred," "five," "ten," and "a dollar." They do not yet understand simple computation skills, but they know greater and lesser amounts.

Level 3

In level 3, children use counting strategies. They have a beginning sense of counting, often using their fingers or objects to solve counting problems. However, they often make counting errors.

Level 4

Students use counting strategies like counting up or counting down when they reach level 4. They find the large number and count up: $8 + 1 = ?$

Level 5

At this level, students have learned number sense. They find correct answers quickly and have learned math computations like $4 + 3 = 7$. They can also perform reverse operations like ($7 - 3 = 4$).

TEACHING BASIC OPERATIONS USING MANIPULATIVE MATERIALS

When students are learning about the operations of addition and subtraction, it is helpful for them to make connections between these processes and the world around them. Story problems help them see the actions of joining and separating. Using manipulative materials and sample word problems gives them experiences in joining sets and figuring the differences between them. By pretending and using concrete materials, learning becomes more meaningful. Tell stories in which the learners pretend to be animals or things. Representing ideas and connecting them to mathematics are the bases for understanding.

When students explore with manipulative materials, they have the opportunity to see mathematical relationships, using the concrete and visual models to help develop their understanding.

Manipulative materials have been used by teachers and students for more than thirty years. A typical elementary classroom has several sets of manipulative materials to improve computational skills and make learning more enjoyable. These materials include base ten blocks, pattern blocks, Cuisenaire rods, geoboards, unifix cubes, and tangrams (Burns, 1998).

Introduce Base Ten Blocks

Base ten blocks will be used in these activities to represent the sequence of moving from concrete manipulations to the abstract algorithms. Students need many chances to become familiar with the blocks, discovering the vocabulary (ones = units, tens = longs, hundreds = flats) and the relationships among the pieces. The following activities will explore trading relationships in addition, subtraction, multiplication, and division.

Activity 1: The Tens Game

In this activity, small groups of students will be involved in representing tens. The game works best by dividing the class into small groups (four or five players and one banker). Each player begins with a playing board divided into units, longs, and flats (units represent ones, longs represent tens, a flat is one hundred). Before beginning, the teacher should explain the use of the board. Any blocks the student receives should be placed on the board in the column that has the same shape at the top. A student begins the game by rolling a die and asking the banker for the number of blocks rolled in *units*. They are then placed in the units column on the student's board. Each student is in charge of checking his or her board to decide if a trade is possible.

The trading rule states that no player may have more than nine objects in any column at the end of his or her turn. If he or she has more than nine, the player must gather them together and go to the banker and make a trade (for example, ten units for one long). Play does not proceed to the next player until all the trades have been made. The winner is the first player to earn five longs. This game can be modified by using two dice and increasing the winning amount.

Activity 2: Regrouping Tens

This game is simply the reverse of the Tens Game. The emphasis here is on regrouping the tens. Players must give back to the bank in units whatever is rolled on the die. To begin, all players place the same number of blocks on their boards. Exchanges must be made with the banker. Rules quickly are made by the students (for example, when rolling a six, a player may hand the banker a long and ask for four units back). It is helpful for students to explain their reasoning to one another. The winner is the first to have an empty playing board. Students should decide in their group beforehand whether an exact roll is necessary to go out or not.

Activity 3: Learning about Division

Base ten blocks help the understanding of an often confusing mathematics process. The following activity is a good place to start when introducing and representing division.

1. Using base ten blocks, have students show 483 with flats, rods, and units.
2. Next, give them a problem: $483 \div 3 = ?$
3. Students divide the blocks into three parts and come up with an answer.
4. Students soon learn how to record their work.

ACTIVITIES

A few innovative activities are presented as a way to connect the mathematics standards to classroom practice. The intent is not to suggest an activity for a particular grade level, but to offer sample activities that could be enhanced and used in many grades:

1. Pair numbers with objects. When discussing objects in the classroom or pictures in a storybook, have students put numbers and objects together. Students begin to associate numbers with different values (Gurganus, 2004). A real-life example: there are two wheels on a bicycle, three wheels on a tricycle, and four wheels on a car. How many wheels are in your garage right now?

2. Begin math class with counting practice. Even young students can usually count to ten. Counting out loud emphasizes counting without the embarrassment of making mistakes.

3. Use number patterns to help counting continue. Students count from 100 to 110, from 200 to 210, from 300 to 310, and so on. Continue counting patterns as long as students are interested.

4. Encourage students to use fingers to help represent numbers. Games and songs using fingers also help students count. "This old man he played one . . ." is one example. Students sing the song using fingers as they count to ten.

5. Examine "less than" and "more than" with students. Discuss estimation examples using "less than," "more than," and "equal to." "I have ten stickers. Bob has six—tell me who has more."

6. Counting off. Students line up for recess, lunch, and just about everything else. So encourage them to count off and use counting patterns orally as they form a line.

7. Be alert to numbers in many situations. (For example, numbers on the clock, numbers in books, numbers in class, etc.) Have students raise their hand when they hear or see the desired number. "Today's number is eight." Students who raise their hands need to explain where the number is located.

8. Model enjoyment of numbers. Curiosity is sparked when teachers are excited about numbers and number games. Play number games with your students.

9. Connect a "sense of numbers" in the students' personal lives. To help students understand the role of numbers in their personal lives, have them do an illustration like the one shown in figure 3.1. They can choose anything from a flower to a bird, as long as they can put the numbers about their personal life in the drawing. This works really well with cardstock paper or it can be done with large or small sheets of construction paper. Although each illustration is about an individual, learners often enjoy working on this differentiated activity with a partner, each doing his or her own personal illustration.

Figure 3.1.   Lourdes's Sense of Numbers

DIFFERENTIATED TEACHING IDEAS

In addition to the teaching strategies mentioned above, other mathematical games and differentiated activities can also assist in developing number sense. These creative ideas should be looked at as mathematics play activities.

Number Sense and Operations

Math Play Activities

1. Catch the Teacher If You Can. All of us make mistakes. Your job is to find any counting errors the teacher makes. The teacher may skip count or repeat numbers while counting. If you hear a mistake quietly raise your hand. You must describe the mistake the teacher made.

2. Traveling Down the Road. Using a number line on the floor, have students move down the road by saying the numbers as they pass by. Older students can use positive and negative numbers.

3. Click Together Objects. Kindergarteners and first graders need to practice addition and subtraction often in different game formats. Use locking cubes or snap-together objects to play the game. Divide the students into groups of two. Give each group a chain of ten snap-together objects. One student breaks the chain and gives the other student the portion of the chain that's left. The second student tells what's missing. Middle school students can practice counting by using variables to represent numbers.

ALGEBRA STANDARD

All students should understand patterns, relations, and functions.

- Students will sort, classify, and order objects by size and number.
- Students will recognize, describe, and extend patterns.
- Students will represent and analyze patterns.

Pattern Blocks to Teach Algebra

Pattern blocks are a collection of six shapes and colors: green triangles, orange squares, blue parallelograms, tan rhombuses, red trapezoids, and yellow hexagons. The shapes have sides that are the same length, except the trapezoid, which has one side that is twice as long. All shapes fit together, which provides a range of explorations.

Pattern Block Activities

1. Sort All the Shapes. (Green triangles, blue parallelograms, red trapezoids, yellow hexagons)

> Sort all the rectangles, all the triangles, all the blue shapes, and square shapes.
> Sort all the four-sided shapes (blue rectangles).
> Sort all the three-sided shapes (green triangles).

2. Make a Pattern Using the Blocks, For example: two triangles, one rectangle. Continue the pattern.

3. Free Exploration. Allow students time to explore with the blocks. They create scenes, count objects, and make a variety of patterns.

Sound Patterns

Sound patterns are things all of us listen to every day whether listening to music or the traffic passing by. The snapping and crackling sounds are everywhere.

4. Have Students Create a Sound Pattern. Have students clap their hands, snap their fingers, hit their desk, and so on. Have them record what they do (x stands for claps, s for snaps, h for hits, etc.).

MEASUREMENT AND ESTIMATION STANDARD

All students should understand length, weight, area, and time. Estimation is an important part of measurement. Often, students are first asked to estimate a measure and then check by measuring. Estimating helps students to understand measurement ideas.

Activity 1: Telling Time Movement Activity

The teacher should design a large clock face (six feet in diameter) on the floor using masking tape to make the circle and placing numbers

1 through 12 in the appropriate spots. A yardstick can be used as the minute hand and a ruler as the hour hand. Students working in teams of two or three should display various times on the clock by moving the hands as necessary. They may be told: "Make the clock say 1:15." To involve all students, each student should make a small clock with movable "hands" and individually complete the same task as a team of students working on the large clock.

Have students sing the telling-time song to the tune of "The Wheels on the Bus Go Round and Round."

"Telling Time"
The short hand says its number first, number first,
When telling time,
The long hand is tall and counts by five,
Counts by five, counts by five,
When telling time.

Students need to be able to count by fives because much of telling time involves increments of five minutes. Practice telling-time using the clocks and singing the telling-time song.

ESTIMATION LESSONS

In grades K–4, the curriculum includes estimation so students can

- Explore estimation strategies.
- Recognize when an estimate is appropriate.
- Determine the reasonableness of results.
- Apply estimation in working with quantities, measurement, computation, and problem solving.

Math instruction in the primary grades tries to make classifying and using numerals an essential part of the classroom experience. Learners need many opportunities to identify quantities and see relationships between objects. Students count and write numerals. When developing beginning concepts, students need to manipulate concrete materials and relate numbers to problem situations (Findell, Cavanagh, Dacey, & Greenes, 2004).

Individuals benefit by talking, writing, and hearing what others think. In the following activity, students are actively involved in estimating, manipulating objects, counting, verbalizing, writing, and comparing.

Directions:

1. Divide students into small groups (of two or three). Place a similar group of objects in a container for each group, which is color coded. Pass out recording sheets divided into partitions with the color of the container in each box.
2. Have young students examine the container on their desks, estimate how many objects are present, discuss with their group, and write their guess next to the color on the sheet.
3. Next, have the group count the objects and write the number they counted next to the first number. Instruct the students to circle the greater number.
4. Switch cans or move to the next station and repeat the process. A variety of objects (small plastic cats, marbles, paper clips, colored shells, etc.) adds interest and provides real motivation.

Pumpkins: How High? How Heavy? How Far Around?

Objectives: Have the class estimate the size, weight, and circumference of the class pumpkins.

Materials: Unifix cubes, pumpkins, tape measure, weighing scales. (It's important to have enough pumpkins of different sizes so that each small group of students has one. Give each group a tape measure and make sure there are several scales around the room. Even bathroom scales will work.)

Activity 1: Guess the Height of the Pumpkins

Objects: One pumpkin.
Procedure:

1. Put the pumpkin on a desk in the room.
2. Make a chart with these questions and room for group members' estimates: How tall is the pumpkin?

3. Have groups use the unifix cubes on their desks to measure the pumpkins. Have students guess the number of unifix cubes they think will be needed to approximate the height of the pumpkin.
4. Instruct student groups to write their names and their estimates on the poster in the front of the room.
5. Have the groups count the unifix cubes to measure the pumpkins.
6. Lastly, have them write their estimates and how many unifix cubes they used in their portfolios.

Activity 2: Estimating How Heavy the Pumpkins Are

Objects: A weighing scale, one 1-pound book, one 2-pound book, and a pumpkin.
  Procedure:

1. Instruct students to hold the first book to feel how much two pounds are.
2. Now, have them hold the pumpkin. Does it weigh the same as the 2-pound book? More or less? Does it weigh as much as two books? How much would that be? Repeat with the lighter 1-pound book for the last two pumpkins. Ask the same questions.
3. Have groups record their guesses on the class chart.
4. To find out how much the pumpkin weighs, put it gently on the scale and have students tally the number.

  Class follow-up: Groups study the estimate chart to find who is the closest.

Activity 3: Finding the Pumpkin's Circumference

Objects: String, scissors, and a pumpkin.
  Procedure:

1. Instruct groups to cut three strips of string to the length they think would go around the pumpkin's waist.
2. Have groups determine the number of inches of their string estimate.

3. Then, have groups hang their string from the chart under their group's marker.

Class follow-up: Groups actually measure the circumference of the pumpkin. Students compare the accuracy of their estimates by checking the estimate chart.

A STANDARD FOR SOLVING PROBLEMS

Problem solving has been the focus of mathematics education for decades. Problem solving means engaging in a task where the solution is not known. A well-known mathematician named George Polya introduced a four-step plan for solving problems: (1) understand the problem, (2) invent a plan or strategy, (3) follow through with the plan or strategy, and (4) check back and monitor results. See if it makes sense.

Problems can be used for different purposes. The solutions are never the same and there is no right answer because of the range of possibilities. Problem-solving plans include guessing and checking, making a table or chart, drawing a picture, acting out the problem, working backward, devising a simpler problem, searching for patterns, using an equation, applying logic, asking someone for help, forming an organized list, using computer simulation, coming up with original ideas, and taking risks.

Effective teachers model the problem-solving strategies. Modeling might include selecting what strategy to use, deciding what options are possible, and checking on student progress as they go along. Learners catch on quickly if guided through this process. Here are a few sample problem-solving activities.

Introduce Interesting Problems

Bring a problem to the class. Have students draw pictures of what the problem is about or act out the problem. Or have a student read the problem out loud for the class, leaving out the numbers. As students begin to visualize what the problem is about, they will have less difficulty solving it. Students work in small groups when coming up with strategies and solving the problems. Students should write how they solved the problem and discuss and check their answers with other groups in class.

Algorithms, Computers, and Privacy Problems

Google-search algorithm engineers are developing a search engine that second guesses users' needs ahead of time. Right now, it is easy for them to check consumer preferences, Twitter, Facebook, and other data banks to determine what you might want. To do this, they have to know more about your life, your friends, and just about everything else. (This brings up many privacy issues that can be discussed in class.)

An algorithm is a step-by-step procedure or math-influenced formula for solving a problem. Algorithms are used in everything from computer science, to following a recipe for Julia Child's "Boeuf Bourguignon."

There is a constant stream of stories about mathematics-based algorithms and their contribution to artificial intelligence, robotics, and a wide array of digital devices. Still, useful robotic welders and "question-answering" computers (IBM) are one thing, while independent machines that can think and act even a little like humans is quite another.

Jaron Lanier (2010) suggests that it would be more useful to all of us if the time spent on digital stunts was put into developing a really good phrase-based search engine. Lanier, author of *You Are Not a Gadget: A Manifesto*, writes that robot teachers are not very sophisticated. And even though robotic surgical devices are more advanced, they are little more than "high-tech puppetry."

Thinking of interactive digital devices or computer programs as people is a mistake that leads us into a bleak space where the unique human attributes of creative thinking, inquiry, and problem solving are diminished.

Even after Google digitizes most books, arranging it so that writers can live as algorithms within a larger global brain is not part of any near-time future. We have to live with the fact that even the most sophisticated algorithms will not allow our whizbang digital devices to make important decisions for us. It may disappoint some technoenthusiasts, but it will be a very long time before we can even think about shirking our responsibility for moving down a thoughtful path toward an enlightened future.

Adding and Subtracting

To paraphrase Steven Strogatz, it's traditional to teach children subtraction after they have learned addition. That approach seems to make sense because the same facts about numbers are used in both processes, although in reverse. "Borrowing," so important to successful subtraction, is only a bit more confusing than that of "carrying"—its counterpart for addition. If children can cope with 24 + 8, they should be ready for 24 − 8 very soon (Strogatz, 2010). Use the following activity to introduce students to addition and subtraction.

Objectives: In beginning grades, the mathematics curriculum includes concepts of addition and subtraction of whole numbers so that students can

- Show meaning for the operations by modeling and discussing a variety of problem situations.
- Relate the mathematical language and symbolism of basic facts to problem situations and informal language.

When children are learning about the operations of addition and subtraction, it's helpful for them to make connections between these processes and the world around them. Story problems using ideas from science and technology help learners see the actions of joining and separating. Using manipulatives and sample word problems gives students experiences in joining sets and figuring out the differences between them. By pretending and using concrete materials, learning becomes more meaningful.

Directions:

1. Divide students into small groups (two or three students).
2. Tell stories in which the learners can pretend to be animals, plants, other students, or even space creatures.
3. Telling stories is enhanced by having students use unifix cubes or other manipulatives to represent the people, objects, or animals represented in the oral problems.
4. Have students work on construction paper or prepare counting boards on which trees, oceans, trails, houses, space stations, and other things have been drawn.

Mathematics is the study of quality, structure, space, and change. When it comes to learning math, it isn't just something to be completed; it's an expanding landscape to be explored. To be creative requires generating many unique ideas (divergent thinking) and combining those ideas into the best result (convergent thinking). This requires more than spending all your time focusing on one topic or set of topics. If you do that, you won't have the knowledge or the mental agility to do the synthesis, connect the dots, and find the next innovative breakthrough. The point here is that you need exposure to different ideas and subjects to be inventive.

Activity: A Newspaper Hunt for Numbers, and More

For this activity to work well, you need enough of the same newspaper for half the class. The teacher looks through the paper before class and adds ten or twelve items to the list that has something to do with math, science, technology, or another topic of interest. Each student team of two gets a copy of the list and a newspaper.

Find and put down the page number for the following:

something that costs less than two dollars
something that costs more than a thousand dollars
a headline or subtitle that contains a number
an estimate or an exact number
the temperature in Boston
an article without a number in it
(Add more items based on the daily paper.)

After some of the groups are done or nearly finished, call the whole class together and go over the answers. Notice how different groups performed.

Afterward: Have students see how math relates to reading and writing by constructing a brief story using the headlines, subtitles, and pictures that can be brought back to the whole class.

Questions:

What would happen if you took a number out of an article?

Do you like a headline better with or without a number?
What big problem covered has to have numbers in the write-up for
   the problem-solving process to work?

(The teacher could have students rewrite one of the articles that relies on numbers *without using the numbers*. Notice how the meaning
is changed.)

AN INNOVATIVE INSTRUCTIONAL EXAMPLE:
"JUMP START" MATH

In *The End of Ignorance*, mathematician, playwright, and author John
Mighton imagines a world where no child is left behind in any subject.
He has also trained many teachers in methods for discovering and understanding mathematical concepts through problem solving. He calls
the approach JUMP. For more see jumpmath.org/about/mighton.

Here we paraphrase Anne McIlroy's example in Toronto's *Globe
and Mail* newspaper. In a sixth grade classroom Mighton writes the
9 times table on the board and asks students to look for patterns. Students look at the numbers 18, 27, 36, 45, 54, 63, 72, 81, and 90 and
make a few guesses. First guess: the digits in each number add up to
9 (1 + 8; 2 + 7; 3 + 6; 4 + 5; 5 + 4; 6 + 3; 7 + 2; 8 + 1; and 9 + 0 all
equal 9). Within minutes, every student in the class is using the sum of
the digits to see if a number is divisible by 9. If it isn't, they learn to
predict what the remainder will be. Students then practice on several
three-digit numbers.

Since he founded JUMP as an extra tutoring program, Mighton has
collected a lot of anecdotal evidence that helps students who struggle
with mathematics. Does it work in regular classrooms? The program is
still under study and the way the Canadian math curriculum evolves has a
lot to do with the answer. Some detractors think that JUMP might be fine
for weaker students, but will hold some of the top students back.

JUMP is often described as "guided discovery," a method that breaks
things down into small steps so that children can practice and master
basic methods before they move on to the next level. When learning
subtraction, for example, students do a whole page of identifying which
problems require regrouping (or borrowing). This builds confidence

needed to go on to the next step and connects to what brain scientists have discovered about learning—practice leads to mastering small skills, which, in turn, can lead to big jumps in performance.

One problem with the program is that once you move beyond basic arithmetic, things get much more complicated. So if you try some of the JUMP methods, we recommend paying close attention to basic math principles.

Mighton has JUMP resources and publications available on his website. Find curriculum-based materials and publications at jumpmath .org/publications and research reports on the JUMP method at jumpmath.org/program/research/research-initiatives.

ASSESSMENT SUGGESTIONS

- Plan instruction that is informed by past assessment.
- Embed instruction with ongoing (formative) assessment.
- Closely watch learning outcomes, problems, and motivation.
- Plan the next instructional steps based on incoming information.
- Assess in a way that helps you figure out how to put together the next lesson.

CONSTRUCTING MEANING IN MATHEMATICS

A constructive, active approach to learning is the best way to stimulate students' interest. This includes building on learners' past experiences and encouraging them to discover their own ideas. This means that students have many ways to interpret mathematics ideas and construct understandings for themselves. Students will be made to participate in math problem solving, science investigations, and projects that engage thinking and reasoning. Working collaboratively in a group situation helps reinvent and reinforce thinking. Students brainstorm together, reveal their ideas, and make sense of the task at hand. Students also assess, reflect on, and evaluate their work.

Some of the effective methods for teaching mathematics in active small group situations include students' writing about how they solved problems, keeping daily logs or journals, and expressing personal attitudes through creative achievements such as building constructions or artwork. A problem rarely becomes completely used up as the understanding of a solution can always be improved upon. Comprehending connections and making thoughtful links are important tools in going beyond the known to see the context, patterns, and relationships.

With the continued emphasis on thinking, collaborating, communicating, and making connections among subjects, students are more in control of their learning. Students will have many experiences with manipulatives, calculators, computers, and real-world applications. There are more chances to make connections and work with peers on interesting problems. The abilities to use basic math understandings and to confidently estimate and check the reasonableness of their estimates are part of innovative learning.

Whether we are making sense of newspaper graphs, identifying the hazards of global warming, or decoding schedules at work, we'll find that mathematics has genuine meaning in our lives. Students need to use the basic facts of arithmetic before they can master the full power of mathematics as they do mental mathematics using the computer, the calculator, or simply paper and pencil. Sadly, learning to do algorithms (the step-by-step procedures used to compute with numbers) alone will not guarantee success with problems that require reasoning ability. The good news is that across the country, the curriculum is upgrading toward making mathematics more innovative and more relevant to today's changing intellectual and societal demands. Teachers have found that students learn more if they have opportunities to describe their own ideas, listen to others, and collaboratively solve problems.

Helping Students Succeed in Mathematics

- Introduce math ideas in real-world settings.
- Teach concepts of the math operations (adding, subtracting, multiplying, dividing).
- Teach students to understand differentiated instruction by tapping into their interests, skills, and learning styles.
- Integrate students' ideas with the mathematics standards.
- Plan interesting and exciting lessons.

- Present innovative ideas on teaching mathematics plus activities and resources.

Education is about inspiration and trusting students to know what to do with important ideas at a very early age. Still, imaginative behavior doesn't occur in a vacuum. The problems in today's world require more than waiting for inspiration to strike. To some degree, every student may be able to generate creative solutions and alternative possibilities. But to bring out creative impulses at school, teachers have to know enough about the characteristics of effective instruction to nurture young minds.

SUMMARY AND CONCLUSION

Promoting creativity and innovation during math instruction doesn't mean neglecting facts. When students express themselves in multiple ways, creativity and factual learning will be increased. As far as innovation itself is concerned, there is a tendency to overplay the short-range effects and underestimate the long-range possibilities.

To give students the confidence to take risks, teachers have to make sure that there is plenty of support when things go wrong. As adult experts know all too well, breakthroughs in mathematics, science, and technology often begin with some kind of failure. Mistakes and even disasters can actually spur innovation.

Imaginative new ideas often get generated in the space between anxiety and boredom. Getting too comfortable can be a problem. Enhancing creativity and innovation has a lot to do with taking ideas or tools from different fields and combining them with mathematics. Technology can help when used wisely. But sometimes, especially out of school, electronic media encourage Americans to hide their heads in the sand of popular culture.

Having students discover their personal voice, strengths, learning styles, and interests is an important portal to generating creativity and innovative behavior. In a differentiated math classroom, teachers stress investigations and problem solving. They use group activities that involve a wide range of students. A central part of the approach is allowing learners a range of choices, giving them the power to make decisions whenever possible. Problem solving, communication, collaborative learning, deductive reasoning, and making connections are all part of what innovative math instruction is about.

Being naive or afraid of mathematics can be a real problem in school, in the workplace, and for citizens in a democracy. It is also a problem if students don't understand how math impacts their day-to-day lives. Experienced math teachers know that each student has a different understanding of math-related issues and that each learning task presents its own set of challenges. They also know that a lot can be accomplished by working on a wide range of problems together.

Everyone can learn math; it's just that not everyone will learn at the same rate or in the same way. The use of collaborative learning teams is a proven and powerful way to motivate individuals and encourage them to use their imaginations as they learn math. Peer support also helps students feel more confident and more willing to take the kind of risks that go hand in hand with creativity and innovation.

Another benefit in linking instruction to the world outside of school is that students will be stimulated to identify problems that they would be interested in solving collaboratively. As teachers go about matching math lessons to the readiness, interests, and talents of individual learners (differentiation), the result is likely to be the development of a natural sense of community in the classroom.

As the standards suggest, mathematics instruction is more than helping students learn how to carry out procedures. Knowing *how* certainly matters, but it is just as important to know *why* mathematics works the way it does. The ultimate goal: helping all students use their imaginations to gain an appreciation of the power, beauty, and fascination of mathematics.

Clearly, when it comes to being well-informed citizens, we all have a responsibility to know something about mathematics. Gaining a deep understanding of the role of numbers in our lives is essential for functioning in today's world. To make this happen for a broad spectrum of citizens requires taking more responsibility for improving our own education and the education of others.

Either directly or indirectly, we are all engaged in the education of children and young adults. And it is time to expect more from ourselves and from students in classrooms around the country.

REFERENCES

Burns, M. (1988). *Mathematics with Manipulatives* [six videotapes]. White Plains, NY: Cuisenaire Company of America.

Findell, C. R., Cavanagh, M., Dacey, L., & Greenes, C. E. (2004). *Navigating through Problem Solving and Reasoning in Grade 1 (Principles and Standards for School Mathematics Navigations).* Reston, VA: National Council of Teachers of Mathematics.

Gurganus, S. (2004). Promote number sense. *Intervention in School and Clinic,* 40(1): 55–58.

Lanier, J. (2010). *You Are Not a Gadget: A Manifesto.* San Francisco: Knopf.

National Council of Teachers of Mathematics (NCTM). (2000, 2010). *Principles and Standards for School Mathematics.* Reston, VA: National Council of Teachers of Mathematics. [NCTM updates this document annually.]

Stephen, M., Bowers, J., Cobb, P., & Gravemeijer, K. (2004). *Supporting Students' Development of Measuring Conceptions: Analyzing Students' Learning in Social Context.* Reston, VA: National Council of Teachers of Mathematics.

Strogatz, S. (2010, February 14). Opinionator: The enemy of my enemy. *New York Times.* Retrieved on July 5, 2010, from opinionator.blogs .nytimes.com/2010/02/14/the-enemy-of-my-enemy/?scp=38&sq=stephen%20 strogatz%20and%20the%20enemy%20of%20my%20enemy&st=cse.

RESOURCES

Adams, D., & Hamm, M. (1994). *New Designs for Teaching and Learning: Promoting Active Learning in Tomorrow's Schools.* San Francisco: Jossey-Bass Publishers.

Athans, S. K., & Devine, D. A. (2010). *Fun-Tastic Activities for Differentiating Comprehension Instruction Grades 2–6.* Newark, DE: International Reading Association.

Barnett-Clarke, C., Ramirez, A. B., & Coggins, D. (2010). *Math Pathways & Pitfalls Percents, Ratios, and Proportions with Algebra Readiness: Lessons and Teaching Manual Grade 6, Grade 7, and Grade 8.* San Francisco: WestEd.

Beckley, P., Marland, H., Compton, A., & Johnston, J. (2011). *Problem Solving, Reasoning and Numeracy (Supporting Development in the Early Years Foundation Stage).* New York: Continuum.

Beckmann, S. (2010). *Mathematics for Elementary Teachers with Activity Manual.* Boston: Addison-Wesley.

Bender, W. (2005). *Differentiated Math Instruction: Strategies That Work for K–8 Classrooms!* Thousand Oaks, CA: Corwin Press.

Benjamin, A. (2003). *Differentiated Instruction: A Guide for Elementary School Teachers.* Larchmont, NY: Eye On Education.

Bennett, Jr., A. B., & Nelson, L. T. (2001). *Mathematics for Elementary Teachers: An Activity Approach.* New York: McGraw-Hill.

Brumbaugh, D. K., Moch, P. L., & Wilkinson, M. E. (2005). *Mathematics Content for Elementary Teachers.* Mahwah, NJ: Lawrence Erlbaum Associates.

Burns, M. (2001). *About Teaching Mathematics: A K–8 Resource.* White Plains, NY: Math Solutions Publication.

D'Amico, J., & Gallaway, K. (2010). *Differentiated Instruction for the Middle School Science Teacher.* San Francisco: Jossey-Bass.

Fisher, R. W. (2010). *Mastering Essential Math Skills Book 1 Grades 4–5 with DVD: Redesigned Library Version.* Peabody, MA: Math Essentials.

———. (2010). *Mastering Essential Math Skills Book 2 Middle Grades/High School New Redesigned Library Version with Companion DVD.* Peabody, MA: Math Essentials.

Fox, S. (2010). *Mathematics across the Curriculum: Problem-Solving, Reasoning and Numeracy in Primary Schools.* New York: Continuum.

Haylock, D. W. (2010). *Mathematics Explained for Primary Teachers.* Thousand Oaks, CA: Sage Publications.

Koshy, V., & Murray, J. (2011). *Unlocking Mathematics Teaching,* 2nd ed. New York: Routledge.

Krech, B., Birrer, D., & DiLorenzo, S. (2010). *SMART Board Lessons: Math Word Problems: Ready-to-Use, Motivating Lessons on CD to Help You Teach Essential Problem-Solving Skills.* Scranton, PA: Scholastic Teaching Resources.

Lee, M., & Miller, M. (2010). *Differentiated Activities for Teaching Key Math Skills: Grades 2–3: 40+ Ready-to-Go Reproducibles That Help Students at Different Skill Levels All Meet the Same Standards.* Scranton, PA: Scholastic Teaching Resources.

———. (2010). *Differentiated Activities for Teaching Key Math Skills: Grades 4–6: 40+ Ready-to-Go Reproducibles That Help Students at Different Skill Levels All Meet the Same Standards.* Scranton, PA: Scholastic Teaching Resources.

Leikin, R., Berman, A., & Koichu, B. (Eds.). (2009). *Creativity in Mathematics and the Education of Gifted Children.* Boston: Sense Publishers.

McIlroy, A. (2010, January15). One small JUMP could become a giant leap in math for regular students. *Globe and Mail.*

McKinney, S., & Hinton, K. V. (2010). *Mathematics in the K–8 Classroom and Library.* Vandalia, OH: Linworth.

Mighton, J. (2008). *The End of Ignorance: Multiplying Our Human Potential.* Toronto, ON: Knopf Canada.

Muschla, J. A., Muschla, G. R., & Muschla, E. (2010). *Math Teacher's Survival Guide: Practical Strategies for New and Experienced Teachers, Grades 5–12.* San Francisco: Jossey-Bass.

Musser, G. L., Peterson, B. E., & Burger, W. F. (2010). *Mathematics for Elementary Teachers, Student Activity Manual: A Contemporary Approach.* Hoboken, NJ: John Wiley.

Nasir, N. S., & Cobb, P. (Eds.). (2007). *Improving Access to Mathematics: Diversity and Equity in the Classroom.* New York: Teachers College, Columbia University.

Onish, L. (2010). *Solve-the-Riddle Math Practice: Addition & Subtraction: 50+ Reproducible Activity Sheets That Help Students Master Addition & Subtraction Skills.* Scranton, PA: Scholastic Teaching Resources.

Posamentier, A. S., & Jaye, D. (2006). *What Successful Math Teachers Do, Grades 6–12: Research-Based Strategies for the Standards-Based Classroom.* Thousand Oaks, CA: Corwin Press.

Pound, L. (2011). *Teaching Mathematics Creatively (Learning to Teach in the Primary School Series).* New York: Routledge.

Powers, W. (2010). *Hamlet's Blackberry: A Practical Philosophy for Building a Good Life in the Digital Age.* New York: HarperCollins.

Ronis, D. (2007). *Brain-Compatible Mathematics.* Thousand Oaks, CA: Corwin Press.

Ryan, J., & Williams, J. (2007). *Children's Mathematics 4–15.* New York: Open University Press.

Scholastic Teaching Resources. (2010). *The Great BIG Book of Funtastic Math: 200+ Super-Fun Activities, Games, and Puzzles That Help Students Master Must-Know Math Skills and Concepts.* Scranton, PA: Scholastic Teaching Resources.

Schulz, K. (2010). *Being Wrong: Adventures in the Margin of Error.* New York: Ecco.

Sinclair, N. (2006). *Mathematics and Beauty: Aesthetic Approaches to Teaching Children.* New York: Teachers College Press, Columbia University.

Starko, A. J. (2010). *Creativity in the Classroom: Schools of Curious Delight.* New York: Routledge.

Teacher Created Resources Staff. (2010). *Ready-Set-Learn: Fractions Grd 3 (Ready Set Learn)*. Westminster, CA: Teacher Created Resources.

Thompson, F. M. (2010). *The Algebra Teacher's Activity-a-Day, Grades 6–12: Over 180 Quick Challenges for Developing Math and Problem-Solving Skills (JB-Ed: 5 Minute FUNdamentals)*. San Francisco: Jossey-Bass.

Tobey, C. R., & Minton, L. (2010). *Uncovering Student Thinking in Mathematics, Grades K–5: 25 Formative Assessment Probes for the Elementary Classroom*. Thousand Oaks, CA: Corwin Press.

Van De Walle, J. (2004). *Elementary and Middle School Mathematics: Thinking Developmentally*, 5th ed. Boston: Pearson Education.

Van Doren, E. (2010). *Get Ready for First Grade: Math & Science*. New York: Black Dog & Leventhal Publishers.

Waterman, S. S. (2009). *Differentiating Assessment in Middle and High School Mathematics and Science*. Larchmont, NY: Eye On Education.

Wells, C. (2010). *Smarter Clicking: School Technology Policies That Work!* Thousand Oaks, CA: Corwin Press.

Whitin, P., & Whitin, D. (2000). *Math Is Language Too: Talking and Writing in the Mathematics Classroom*. Reston, VA: National Council of Teachers of Mathematics; Urbana, IL: National Council of Teachers of English.

Willis, J. (2010). *Learning to Love Math: Teaching Strategies That Change Student Attitudes and Get Results*. Alexandria, VA: Association for Supervision and Curriculum Development.

CHAPTER 4

Science Instruction

Promoting Inquiry and Innovation in the Classroom

Science instruction should be an interesting and exciting process of learning about the world and beyond. Science-related concepts enliven student curiosity, interest, and knowledge. Innovation is a close associate of scientific inquiry. Inquiry skills like observing, experimenting, and questioning are all closely related to the scientific method, creativity, and critical thinking. Many aspects of science (including classifying, measuring, collecting, and experimenting) come naturally to many students. But one size definitely doesn't fit all when it comes to learning how to use scientific processes.

One of the goals of science instruction is to open students' minds and expand their perception and appreciation of the nature of life. Examples include water, rocks, plants, animals, people, and other elements in the world around them. Clearly, active participation in the processes of science is a good way for students to go about learning science and related procedures.

What is the best way to get students to experience and make sense of the science-associated tools, objects, events, ideas, or problems? Good teachers are central to everything, including making sure that learners know what to do when they do not have an answer.

Teaching science requires being familiar with scientific inquiry, the science standards, and the characteristics of effective instruction. As the natural world in which they live is explored, both teachers and students become coaches and players in the world of science.

Science, mathematics, and technology reinforce each other, each drawing from the techniques and tools of the other. To make the interdisciplinary connections clear, scientific processes can be incorporated into other aspects of classroom life. Making it all happen has a lot to do with helping students appreciate these subjects as vital, interrelated elements of their daily lives.

SCIENTIFIC INVESTIGATION AND REASONING

Scientific investigation often winds its way through demonstrations, experiments, and activities where students perform a task to collect data and acquire facts. The process involves observing, asking questions, inquiring, and recording what has been learned. Along the way, students can be encouraged to engage in small group discussions and collaborate. During—or toward the end of each task—teamwork skills are encouraged if learners are asked to compare what was recorded to see if they can reach a consensus.

To help students understand how to conduct a discussion, it's important to ask inquisitive questions.

Questions that begin with phrases such as

"I wonder if . . ."
"How can . . ."
"How come . . ."
"What effects . . ."
"Does . . ."

These beginning ideas help students learn how to ask critical questions through observing and group discussion. Throughout the process, students can learn more through their inquiry activities by working with peers, doing more research, designing, predicting outcomes, and experimenting. Pairs or small groups of students conduct the experiment, collect data, and report on what they find. Finally, they can share what they found with other groups or the whole class. (Posting it online is often a good idea.)

THE CHANGING SCIENCE CURRICULUM

Over the last twenty years, there have been many suggestions for changing the goals of science instruction. Changes have also been taking place in the culture, the schools, and how people live and work. In responding to these conditions, the science standards have updated certain concepts and principles from biology, chemistry, earth science, and physics.

One of the most important changes in science education relates to the changes across scientific fields. Science now pays more attention to finding solutions to personal and social problems. At the same time, the school problems we face are more difficult and involved than ever before. The same can be said for what's happening outside of school on account of a more global economy, an information era in full force, differing family structures, and a knowledge-intensive environment. The whole mix points to a new world of life, learning, and work.

Today's active science curriculum has the power to make a difference in the lives of students and in the society where they live. Some suggestions from the National Science Education Standards (NSES) include

1. Emphasize inquiry skills and knowledge of the subject.
2. Focus on problem solving and collaboration.
3. Make provisions for differentiation.
4. Provide a basic core of subject matter.
5. Coordinate closely with subjects like mathematics and technology.
6. Make the curriculum more relevant to students' lives. (NRC, 1996, 2010)

An important suggestion in the standards is adding greater depth to what is covered in science curriculum. Sometimes less is more; deeply focusing on a smaller number of skills and concepts may lead to greater understanding and retention.

The science curriculum is closely linked to related subjects such as mathematics and technology. Instead of separate subjects, elementary science is often part of an integrated curriculum. The goal is competency and improved attitudes—giving everyone the tools they need to understand science-related issues.

In the twenty-first century, schools around the world are trying to be incubators for scientific and technological innovation. Even in the most developed countries, this is a complex and difficult challenge. In the United States, for example, the task is especially difficult in schools that have large numbers of students who are poor, homeless, or not fluent in English. But even in the direst circumstances, when teachers are given the resources, support, professional development, and a high-quality curriculum, they can make a real difference.

THE EXCITING WORLD OF SCIENCE EDUCATION

In the last century, science was often viewed as a hard subject to teach—with science terms difficult to understand. Too many students felt science was boring and schools were paying more attention to standardized tests than real science. Whatever the reason for difficulty, those who aren't naturally drawn to the sciences need to know that they are not alone.

Science can be the most exciting experience for students and teachers when it is taught as an active, hands-on subject where students learn through doing. Inventive teachers can make twenty-first-century science instruction come alive.

DIFFERENTIATION AND RESPONDING TO EACH STUDENT'S LEARNING NEEDS

Differentiated learning focuses on the idea that teachers adapt instruction to meet student needs. Teachers try to reach all students by providing the right level of challenge for students who perform below grade level, advanced students, and those in between. They are working to deliver instruction in ways that meet the needs of all learners.

The following points are teaching suggestions for differentiated instruction. They are starting points for consideration, not a complete guide. We encourage you to revise and edit the list.

1. Use Collaboration as a Learning Tool.

Collaborative learning lends itself to differentiated instruction. Every student must think, learn, and teach others. Within a collaborative learning classroom, there will be many and varied strengths among students. Each student possesses characteristics that will help to enrich learning for all students. Sometimes it means having the ability to do exceptional work. In the collaborative learning classroom, each student is allowed to work at his or her own level.

Teachers try to "accommodate" students by making changes in the ways that they teach and in the ways they work with students. They form challenging environments open to individual differences. For students, the emphasis is on performance. Teacher collaboration means detailed planning, good teaching, performance assessment, and reflection.

2. Create Changeable Groups.

To ensure each learner's potential, teachers should know their students and meet each student's needs. When they respond to student differences, it benefits all students. Flexibility when grouping students gives students many opportunities to build their strengths and show their performance.

3. Establish Learning Centers.

A learning center is a space in the class that contains a group of activities or materials designed to teach, reinforce, or extend a particular concept. Centers generally focus on an important topic and use materials and activities addressing a wide range of reading levels, learning profiles, and student interests.

A teacher may create many centers such as a science center, a math center, or a reading center. Students don't need to move to all of them at once to achieve competence with a topic or a set of skills. Have students rotate among the centers. Learning centers generally include activities that range from simple to complex.

Effective learning centers usually provide clear directions for students, including what a student should do if he or she completes a task, or what to do if he or she needs help. A record-keeping system should be included to monitor what students do at the center. An ongoing assessment of student growth in the class should be in place, which can lead to teacher adjustments in center tasks.

4. Develop Interesting Activities for All Students.

These are helpful strategies when teachers want to address students with different learning needs. For example, a student who struggles with reading from a science textbook or has a difficult time with complex vocabulary needs some help in trying to make sense of the important ideas in a given chapter. At the same time, a student who is advanced well beyond grade level needs to find a genuine challenge in working with the same concepts.

Teachers use interesting science activities so that all students focus on necessary understandings and skills but at different levels of complexity and abstractness. By keeping the focus of the activities the same but providing different routes of access, the teacher maximizes the likelihood that each student comes away with important skills and is appropriately challenged.

Teachers should select the concepts and skills that will be the focus of the activity for all learners. Using assessments to find out what the students need and creating an interesting activity that will cause learners to use an important skill or understand a key idea is part of a tiered, step-by-step approach. It is important to provide varying materials and activities. Teachers match a version of the task to each student based on student needs and task requirements. The goal is to match the task's degree of difficulty and the students' readiness.

5. Promote Flexible Learning.

Challenging strategies put more emphasis on authentic problems where students are encouraged to formulate their own problems on a topic they're interested in and work with others to solve it. Problems are connected to the "real world" and allow time for discussion and sharing of ideas among students.

6. Use the Science Standard.

Integrating standards into the curriculum helps make learning more meaningful and interesting to reluctant learners. Having a clearly defined set of standards helps teachers concentrate on instruction and makes clear to students the expectations of the class. Students come to understand what is expected and work collaboratively to achieve it. Challenging collaborative groups to help each other succeed is another way to avoid poor performance.

7. Provide Many Learning Possibilities.

Not all students learn in the same way or at the same time. Teachers can provide many learning options by differentiating instruction. This means teachers reaching out to all students to improve teaching in order to create the best learning experience possible.

8. Get Students Involved in Active Reading.

This is an approach that uses "active reading" ideas to improve students' abilities to explain difficult text. This step-by-step process involves reading aloud to themselves or someone else as a way to build science understandings. Although most learners self-explain without verbalizing, this active reading approach is similar to that used by anyone attempting to master new material: the best way to truly learn is to teach and to explain something to someone else.

THE CHANGING SCIENCE CURRICULUM

Today, active science learning in the elementary and middle schools is changing the boring textbook process. It contributes to the development of interdisciplinary skills. For example, the overlap in science and mathematics is obvious when you look at common skills. Many of the best models in science education involve having students work in cross-subject and mixed-ability teams. Teachers begin by making connections among science, mathematics, and real-world concerns (a good example would be those found in the newspaper). The live action of science education and literacy are in the hands of teachers.

To use and understand science today requires an awareness of what the scientific endeavor is and how it relates to our culture and our lives. Inquiry involves curiosity, observation, posing questions, and actively seeking answers.

THE SUBJECT MATTER STANDARDS FOR SCIENCE

The content standards suggest what science learners should know, understand, and have the ability to do. Students should be able to

- Comprehend the basic skills, ideas, and processes in science.
- Apply the inquiry process when doing science.
- Study physical science, life science, earth science, and space science when doing activity-based learning.
- Apply science understandings when designing solutions to problems.
- Make the connections among science, math, and technology.
- Practice science from personal and social viewpoints.
- Become familiar with the history and nature of science through readings, observations, and writings (NRC, 2000, 2007).

Inquiry and the Science Standards

The inquiry skills of science are gained through a questioning process. Inquiry forms new questions and provides new directions to examine. The findings may generate suggestions and point to different ways of expressing ideas. The inquiry process helps students with science content knowledge and skills. It also invites learners to explore anything that interests them. Whatever the problem, subject, issue, or thoughtful question, the inquiry process will use some of the same thinking processes that are used by scientists who are searching for new knowledge in their respective fields of study.

Inquiry processes form a foundation of understanding and are components of the basic goals and standards of science and mathematics. These goals are intertwined and multidisciplinary, providing students many opportunities to become involved in inquiry. Each goal involves one or more processes (or investigations). The inquiry process approach includes the major process skills and standards as outlined in the activities that follow.

The science activities also include the key principles of a differentiated classroom. This includes the **content**—what students will learn; the **process**—the activities by which students make sense of important ideas using necessary skills; the **product**—how students show what they have learned and prove their point; and the **learning environment**—safe, comfortable conditions that set the tone for learning (Tomlinson, Brimijoin, & Narvaez, 2008).

The Science Standards and Science Activities

This section links the science standards to the elementary and middle school classrooms. It establishes activities that use the inquiry skills of observing, measuring, recording data, and drawing reasonable conclusions. Whenever possible, mathematics is included in the activities so that math and science skills are developed together. At the end of each activity, suggestions for differentiated instruction are offered. These ideas provide "a peek" into the differentiated process so that teachers can try out some differentiated strategies with their students.

ACTIVITIES AND LESSON PLANS FOR SCIENCE

Lesson 1: The Science of Yo-Yos

Topic of lesson: Science of yo-yos.

What do you want students to learn? Physics of yo-yo manipulation. Students will observe concepts about force, acceleration, friction, and gravity.

Why are the concepts important? Students can apply science to real-world situations.

What background information do students need before starting? Students should examine the principles of gravity and review basic physics concepts of inertia.

Organization and procedures: This activity will take one or two class periods. Pairs of students will experiment with yo-yos. Students will try to maneuver yo-yos in different ways.

Materials: yo-yos, science and math journal, pens.

How are you going to get the students involved? Students will work in pairs to play with their yo-yos. Hands-on involvement starts after instruction about how yo-yos work. The lesson will challenge students to try to show how gravity works.

Lesson development, questions, and desired product: Allow students time to play with yo-yos, working in pairs. Discuss what they did during their free play time—how they maneuvered their yo-yos and what tricks they tried. Share different physics concepts that influence motion, speed, and direction.

Small group options: Groups of two work well with this activity.

Gearing up (if the lesson is too easy): Have students estimate the different directions that their yo-yos moved during a period of time.

1. Record the velocity (or rate of change of its position). Guess the speed at which the yo-yo traveled.
2. What are other objects that you know that travel at a similar rate of speed?
3. If you were to travel in a car from your house to school, how many turns would you make?
4. Draw a map of your car trip recording all turns.

5. Can you think of an imaginary space vehicle that could travel at a similar rate of speed as the yo-yo? Write a short story about it.
6. Write a short paragraph of how yo-yos demonstrate how gravity works.
7. Give some examples of gravity.
8. Why does a yo-yo behave as it does?
9. There are some professional yo-yo groups. Go online to discover more about them.
10. Record your findings.

Gearing down (if the lesson is too hard): Have students slow down the yo-yo and then try to speed it up. Record the movements in their journals.

1. Count the number of times your yo-yo went up and down.
2. Describe the yo-yo motions you see.
3. What causes the yo-yo to move?
4. What directions does the yo-yo move when it is dropped?
5. What happens when the yo-yo is thrown in a different direction?
6. When does your yo-yo travel the fastest?
7. What makes the yo-yo slow down?
8. What might make the yo-yo move slower?
9. Record the number of times your yo-yo went up and down when you slowed your yo-yo down.
10. What happened when you made your yo-yo go faster?
11. Compare your results with those of your partner.
12. Make a group chart showing the slow and fast movements of your yo-yos.

Assessment (observations, products produced, and portfolio entry):

1. Have students write and describe the physics involved with yo-yos in their portfolios.
2. With partners, ask students to describe the changing motions of a yo-yo using the vocabulary words: *force*, *speed*, *gravity*, and *friction*.

3. Allow partners to help each other clarify explanations as they practice.

4. Record on a class chart which students are able to successfully use the vocabulary in their explanation.

5. Have each student design a poster including diagrams to illustrate the motion of the yo-yo as it falls and moves back up the string.

6. Ask students to identify where the yo-yo moves fastest and slowest with labels and arrows. Use the vocabulary words in their descriptions.

Extensions: Have a yo-yo design contest. Students can learn more about modern yo-yo design by doing web searches on these topics:

How Yo-Yos Work
Reinventing the Yo-Yo

Lesson 2: Comparing Size and Weight

Topic of lesson: Comparing size and weight.

What do you want students to learn? Students should understand the correlations among density, weight, and size.

Why are the concepts important? Students learn that two objects that are of the same size might have different weights. This activity introduces basic math and science concepts.

What background information do students need before starting? Students need to be able to read numbers. Students need to know what each object feels like and looks like. Students need to know what a scale is and how it works.

Organization and procedures: Give small groups of students three or four candies that are similar in size but different in weight. Ask the children to guess which would be heavier. Then have students weigh the candies to see if they guessed right.

Materials: three to four candies of similar size and different weights, scales, wall charts to record findings.

How are you going to get the students involved? This is a hands-on, small group activity. Students will be actively involved in measuring. Candy is a motivator that reaches most students.

Lesson development, questions, and desired product: Students will be involved in the science and math skills of observing, comparing, measuring, estimating, and recording data. Which candy is heaviest? Lightest? How can you prove it?

Small group options: Students will work in small groups to observe and estimate the weight of several candies. Then children will weigh the candy and record their findings on a chart.

Gearing up (if the lesson is too easy):

1. Estimate the weight of the candies.
2. How many of the lighter candies would equal the heaviest?
3. Record the weight of the candy (lightest to heaviest).
4. What is the difference in weight between the lightest and the heaviest?
5. Find the difference between what you estimated and the actual weight.
6. Measure each candy's length, height, and weight.

Gearing down (if the lesson is too hard):

1. Estimate and record the weight of the smallest candy.
2. Weigh the smallest candy and record its weight.
3. Estimate the weight of the largest candy and record it as well.
4. Weigh the largest candy. Record its weight.
5. Compare the two candies. Record how much they weigh.
6. Taste the candies.

Assessment (observations, products produced, and portfolio entries):

1. Make a chart showing the weight of the candies.
2. Write a sentence about your candy.
3. Tell a story about the two candies.
4. Put your results in your portfolio

Lesson 3: Dental Hygiene

Topic of lesson: Tooth health.

What do you want students to learn? The importance of properly brushing and flossing your teeth and the effects of bacteria on teeth.

Why are the concepts important? Students learn about proper techniques for healthier hygiene results. Students become familiar with the anatomy of the mouth.

What background information do students need before starting? Students need to recognize that cleanliness prevents infections and illness.

Organization and procedures:

1. Students will be shown slides of bad and good teeth.
2. Teachers will explain the basic science behind plaque and gum disease.
3. Teacher demonstrates good cleaning habits.
4. Students practice brushing and flossing.

Materials: disposable toothbrushes (can get from dental office, usually provided free), pocket mirrors to look at teeth, model sculpture of teeth (for teacher).

How are you going to get the students involved? This is a hands-on activity. Students will be actively involved in finding out about tooth hygiene and applying directions for teeth and gum cleaning.

Lesson development, questions, and desired product: Explain teeth health step by step. Also, explain that eating involves chemicals breaking down food; that saliva is part of the process; and that plaque is a substance that accumulates between teeth and can lead to cavities. Bacteria are microscopic but often can form in cavities. Explain steps in proper brushing and flossing to avoid plaque and cavities.

Small group options: Students can practice with each other using toothbrushes and pocket mirrors.

Gearing up (if the lesson is too easy):

1. Extension of student vocabulary (dental hygiene, bacteria).
2. Create a chart showing the steps of tooth decay.
3. Write a letter explaining how to take care of your teeth.
4. Use creative drama showing what happens when cavities form.
5. Practice flossing and brushing; explain the steps you used.
6. Suggest healthy eating habits that are tooth friendly.

7. Design a handout showing the directions for teeth and gum cleaning.

Gearing down (if the lesson is too hard):

1. Have pairs of students practice brushing and flossing.
2. Use pocket mirrors to observe your teeth.
3. Keep a record of when you brush and floss your teeth.

Assessment (observations, products produced, and portfolio entries):

1. Teacher observation is a good assessment strategy for this lesson.
2. Have students write about teeth health in their portfolio.

Lesson 4: Exploring Polyhedrons

Topic of lesson: Polyhedrons.

What do you want students to learn? Spatial reasoning—how to construct polyhedrons using manipulatives.

Why are the concepts important? Students can apply this concept to more advanced geometry.

What background information do students need before starting? Students learn the geometric term *polyhedron* and review shape recognition.

Organization and procedures: Have students work in groups of two. Pass out materials and problem sheet directions.

Materials: plastic click-together polyhedrons, triangles, squares, pentagons, and hexagons; toothpicks or straws and marshmallows; science and math journal.

How are you going to get the students involved? Encourage students to explore plastic materials on their own. Then have students work together to make "closed" shapes such as pentagons, polyhedrons, triangles, squares, and hexagons.

Lesson development, questions, and desired product: Start with free exploration, and then move to guided instruction of shapes. How many sides, vertices, and faces does each figure have? Have students describe how many different closed shapes they made. Then have students draw and label them (edge, face, vortex).

Small group options: Small groups of two to four work well with this activity.

Gearing up (if the lesson is too easy):

1. Work together to combine different shapes and make a giant closed shape. Invent a name for it.
2. Label all your shapes.
3. Using toothpicks and small marshmallows, form a pentagon. (A pentagon is a shape that has five sides that are all the same length.)
4. Make eleven more pentagons.
5. Place one pentagon on the table and attach five pentagons around it. You're in the process of building a geodesic dome.
6. Finish your structure. Write about it.

Gearing down (if the lesson is too hard):

1. Make simple shapes and label them.
2. Describe the shapes you made. (How many sides, how long, how high?)

Assessment (observations, products produced, and portfolio entries):

1. Teacher observation is a good assessment strategy for this lesson.
2. Observe ease of use and understanding.
3. Ask students questions.
4. Have students write and define shapes in their portfolios.

Lesson 5: Fall Fruits

Topic of lesson: Three fall fruits.

What do you want students to learn? How to classify different fruits (same and different) and compare and contrast each fruit.

Why are the concepts important? They help students sort facts and organize information.

What background information do students need before starting? Students need to have eaten, seen, or touched pumpkins, apples, and oranges.

Organization and procedures: Give each group a pumpkin, an apple, and an orange. Ask them to describe each fruit. How are they alike? How are they different?

Materials: poster paper, markers and crayons, thinking cap.

How are you going to get the students involved?

1. Bring in the three fruits—apples, oranges, and pumpkins.
2. Encourage students to compare the fruits by drawing a picture about them.
3. Ask students to look at things that are similar about the fruits.
4. Next, have students list the differences.
5. Students can then make a chart of what they found.

Lesson development, questions, and desired product: To understand how apples, oranges, and pumpkins are different but similar.

Small group options: Two people in a group working together.

Gearing up (if the lesson is too easy):

1. Encourage students to compare the apples, oranges, and pumpkins with other fruits such as bananas, pears, and grapes.
2. Write how they were alike, and how they were different.
3. Add a strawberry or oddly shaped fruit such as a pineapple.

Gearing down (if the lesson is too hard): Use only two fruits to compare.

Assessment (observations, products produced, portfolio entry):

1. Students can make a fun diagram based on the fall fruits.
2. Areas such as nutrition, math, and science concepts fit in easily.
3. Students can also look at fall vegetables and make comparisons.
4. Teacher *observation* is a good way to assess this activity.
5. Have students write and classify the fruits in their portfolios.

Lesson 6: Candy Skittles Square

Topic of lesson: Skittles square. (This is a cooperative group model.)

What do you want students to learn? Students will guess how many Skittles can fit in a 6-inch square. Students will estimate the number of Skittles in a square yard, square mile, and football field.

Why are the concepts important? It enables students to build and reinforce estimation skills. Students will learn how to solve a problem using a formula for finding the area of a square.

What background information do students need before starting? Students should know how to calculate how many Skittles will fit inside progressively larger square units using a formula for finding the area of a square.

Organization and procedures:

1. Pass out 12-inch square pieces of paper, worksheets, and Skittles candy.
2. Organize students so they are in four groups.
3. Ask students to guess how many Skittles will fit in a 6-inch square.
4. The teacher explains how to find out how many Skittles will fit in a 6-inch square using the formula for area of a square ($s \times s$ or s^2).
5. First, see how many Skittles will fit along two sides of the square. Since a square has equal sides, the formula for area is $s \times s$, or $12 \times 12 = 144$ Skittles in this 6-inch square.

Materials: 12-inch squares of brightly colored card stock paper with a 6-inch square marked off, Skittles candy, worksheet with questions, data forms.

How are you going to get the students involved? Interesting questions using candy to find the answer.

Gearing up (if the lesson is too easy):

1. Have students calculate how many Skittles in a square yard.
2. For more challenging questions:

"How many Skittles in a square mile?"
"How many Skittles in a 48,000-square-foot football field?"

Gearing down (if the lesson is too hard): We know there are 144 Skittles in a 6-inch square, how many are in a square foot? (A square foot is 12 inches on all four sides and 24 Skittles can be placed along each side, so the area is 24 × 24 = 576.)

Assessment: Have students pass in their worksheets and check for understanding.

Lesson 7: Convection

Topic of lesson: Thermal energy and free convection.

What do you want students to learn? Students will gain an understanding of thermal energy and convection through the search for an ideal place for their class pet hamster. Students will record measurements about the temperature and air movements in the classroom.

Why are the concepts important? The basic idea behind free convection is that warmer air becomes more buoyant and "rises," while cooler air "sinks."

What background information do students need before starting? Students should know that fluid cools by losing heat from the surface. In a convection cell, warm, low-density fluid rises, while cool, high-density fluid sinks.

Science facts:

1. Heat can be transferred by radiation, conduction, and convection.
2. Thermal energy moves from one place to another because of the difference in temperature.
3. Thermal energy can be transferred from one body, usually hotter, to a second body, usually colder, by conduction.

Materials: four thermometers, four tinfoil spirals, recording books, chart paper and markers, class hamster.

Organization and procedures:

1. Have students identify their group's quadrant of the room. Spend two to three minutes discussing the relative advantages/ disadvantages of their quadrant as a future location for the hamster's cage. Students may make predictions about the temperature and air movement patterns.

2. Identify the tools for data collection: thermometer and tinfoil spiral. Go to the quadrant and test the tools and gather data.
3. Make a sketch of the placement of tools in the quadrant. Record the current data in the class recording book and include any other observations.

How are you going to get the students involved? The survival of the class pet hamster will hopefully be engaging and interesting to all students.

Lesson development, questions, and desired product: The activity will get students to invest in the care and responsibility of the class pet. Students will make a chart showing the perfect environment for the pet hamster.

Gearing up: Ask students how they think energy moves. (Heat energy escapes or spreads out, due to differences in temperature of the surrounding area.) Where does heat energy go once it has left the object?

1. Students investigate the difference between heat (thermal energy) and temperature by having two different containers of water. Both start off at the same temperature. Measure the temperature decrease. Larger containers should take longer to cool (more particles give it more thermal energy so it retains temperature longer.)
2. Try it. Explain conduction in terms of particles, giving examples of each.
3. Write a newspaper article suggesting why people should use renewable energy resources whenever possible.

Gearing down: Have students measure the temperature of four parts of the room. Students will learn about the movement of air through actively exploring their class environment. Students will be completing activities with their group.

Assessment: Student understanding will be checked by examining records.

Lesson 8: Science and Math Algebra Poker

Topic of lesson: Algebra. The game of algebra poker is designed to reinforce or teach missing number problems in any given operation. Examples of missing number operations reinforced:

$3 + x = 10$ answer is 7
$4 \times x = 32$ answer is 8

Materials: deck of cards for each group, optional worksheets for written reinforcements of problems solved.
Procedures:

1. Students form groups of three. Two students are players, one is the judge.
2. Deck of cards is shuffled. Two students draw randomly and hold card face out so they are unable to see their own card.
3. The judge announces the answer to the math problem using numbers on both cards (first announcing the operation being used: "we're doing addition right now").
4. As each player knows what the other card is, they are able to guess their own card by working out the math problem.

Jenny has the card with a 7 up against her forehead. Scott has a 4. The judge announces that the answer is 11. Scott can figure that out because Jenny has a 7, so he must have a 4. The first student to guess his or her own card wins.
Structure:

1. Introduce class to the missing number operations game.
2. Model the game in front of the class.
3. Have students form groups.

Gearing up: Change the operation. The game can be done with multiplication and even equivalent fractions. Can you invent other ways to use it?
Gearing down: The game can be done with addition or subtraction.
Assessment:

1. Students' understanding will be checked by the work done in groups and by fellow students.
2. Students collect the cards at the end of the each match. The student with the most cards wins.

3. Have students reflect on the game. What was exciting about it?

4. Did it improve your performance?

Lesson 9: Taking Care of the Earth

Topic of lesson: Environmental awareness. Math and science activities are offered that may help students better understand how their presence and actions can and do impact our world. Did you know that on average each person throws away about 4.4 pounds of trash every day?

WORD Problem: Quick Quiz

1. On average, how much does each person throw away in a week? *(30.8 lbs)*
2. On average, how long will it take for each person to throw away 100 pounds of garbage? *(22.7 days)*
3. On average, how much garbage will a person throw away this year? *(1,601.6 lbs)*
4. At this rate, would a person your age have contributed a ton of garbage? On average, how long does it take for each person to throw away a ton, or 2,000 pounds of garbage? *(22.7 days)*
5. So far in your lifetime, about how much garbage have you contributed?

Gearing up: Estimate what it would take to reduce the garbage in this class, in this city, and in this country. Write a letter to the school newspaper voicing your views.

Gearing down: Landfills in the United States have charged between $10 and $100 per ton to dump trash. If it costs $20 per ton, estimate how much money will be spent this year?

Assessment: Students' understanding will be based on their quiz answers.

Lesson 10: Subtracting with Science and Math Regrouping

Objective: The students should be able to regroup subtraction with math and science problems without any guidance from the teacher.

Theme and/or motivation: The students will be working with base ten blocks and have a rhyme to help them remember the rules for regrouping problems.

Materials: base ten blocks, subtraction rhyme, math worksheet.

Launching the lesson:

1. Divide the class into groups of three or four.
2. Start off by handing out the math worksheet with the rhyme written on the top.
3. Read the rhyme to the class:

   More on top?
   No need to stop.
   More on the floor?
   Go next door.
   Get ten more.
   Numbers the same
   Zero's the game.

4. Have the class read the rhyme together. Do an example of each stanza on the chalkboard.
5. Pass out the base ten blocks.
6. Do the first problem together as a class using the base ten blocks and saying the rhyme if needed.
7. Continue with the problems as a class until students feel confident that they can do the problems on their own.

Class group/individual activities: If at any time students feel that they can go ahead, let them and have their group members check it over. If a student completes the worksheet ahead of everyone else, have him or her help out a student that might not be getting the concept. This activity can be done in a whole class setting or in small groups.

Providing for different interests, needs, and aptitudes: This lesson provides several different strategies for the students to understand the concept and its different aspects. There is a *musical aspect*. Sing the rhyme to the tune of "Jingle Bells." Explore the *visual aspect* using the tens and ones model. Illustrate the *auditory aspect* using rhyming words. Show the *kinesthetic aspect* by having students model what they're doing with the blocks.

Evaluation: Through the worksheet that should be completed at the end of the lesson, students and teachers should have a sense of each student's understanding of regrouping.

Lesson 11: Science and Math Activity: Having Fun Rounding Numbers

Topic of lesson: Rounding whole numbers and decimals.

Objective: Students will round numbers to the nearest specified place value (ones, tens, hundreds, thousands, tenths, hundredths, thousandths).

Theme: Understanding place value using fifth grade text.

Materials: class set of graph paper with larger than usual squares, pencils, fifth grade textbook, yardstick, tape.

This activity looks at water pressure, gravity, volume, weight, and ways to solve problems. The teacher differentiates by making it clear what students are to learn. He or she should understand, appreciate, and build on student differences, adjusting content, process, and product in response to student readiness, interests, and learning profile.

Lesson 12: Bird Identification Jeopardy

Topic of lesson: This lesson is taught in succession with the life science unit called "Habitats."

Source: Adapted from the game of Jeopardy.

Procedures:

1. There will be five stations providing information that students will visit. Reading and pictorial information are provided. Students will move from station to station in groups. Individuals gather as much information as they can about the birds and their habitats by reading and observing only.
2. The students are not to talk and not to take notes. The students will then gather information and "put their heads together."
3. Students will choose a speaker that will answer questions aloud.
4. The group must come to a decision on one answer.
5. Teachers will read the questions that are prepared in a Jeopardy fashion.

Lesson description: This lesson includes opportunities for English learners to work together with other students in a fun, facilitated activity. There are pictures representing the information needed for the game.

Standards: Students will learn to identify five species of local birds and their habitats.

Life sciences: Plants and animals meet their needs in different ways.

1. Students know that different plants and animals inhabit different kinds of environments and have external features that help them thrive in different places.
2. Students know that both plants and animals need water, animals need food, and plants need light.
3. Students know that animals eat plants or other animals for food and may also use plants or even other animals for shelter and nesting.
4. Students know how to infer what animals eat from the shapes of their teeth (e.g., sharp teeth: eats meat; flat teeth: eats plants).
5. Adaptations may improve an organism's chance for survival.

Materials: large poster board, 3 × 5 cards, photos of birds, photos of habitats, written information, individual bird displays (American robin, Anna's hummingbird, red-tailed hawk, Steller's jay, turkey vulture, bar-tailed godwit).

Procedures: Students are welcomed to the Jeopardy game. The teacher explains that this memory concentration is actually a test; students have only twenty minutes to complete the game and determine who among them can remember things about these five birds.

1. Students have about one minute per station.
2. Students will be asked to move from station to station in groups.
3. Students are asked not to talk and not to take notes.
4. Students gather together and collaborate.
5. Students choose a speaker to answer questions aloud.
6. The group comes to a decision on one answer.

Written information:

American Robin: The robin is eight to eleven inches long and has a wingspan of twelve to sixteen inches. The male robin has gray or brown back and wing feathers and reddish-orange chest feathers. The female has the same color pattern, but she is a bit duller in color. The robin eats a wide variety of foods, including fruits and berries, worms, grubs, and caterpillars. The robin lives in open woodlands, fields, gardens, and yards.

Anna's Hummingbird: Historically, this bird is limited to western California; in recent years, however, this class of birds expanded its territory. Males and females do not form lasting pair bonds. Females construct the nest, incubate the eggs, and feed the nestlings on their own. Hummingbirds feed on nectar from flowers and feeders as well as on the spiders they catch in the air or on tree trunks and branches.

Turkey Vulture: The turkey vulture's head is bald and red. The turkey vulture does not feed strictly on carrion. The bird enjoys plant matter as well, including shoreline vegetation, pumpkin, and bits of other crops. The turkey vulture soars above the ground for most of the day searching for food with its excellent eyesight and highly developed sense of smell. Size: twenty-five to thirty-two inches long with a wingspan around six feet. A healthy adult vulture weighs about six pounds.

Steller's Jay: Front half of body is sooty black, rear is vibrant blue, with a charcoal-colored head and nape with a large black crest on top of head and black cross barrings on wings and tail. Harsh metallic voice sounds chack, chack, chack. Steller's jays are omnivores, and their diets are about two-thirds vegetable matter and one-third animal matter. The vegetable portion of their diet consists of seeds, nuts, berries, and fruits, and the animal portion consists of bird eggs, small rodents, reptiles, and carrion. They eat scraps that humans give them. Steller's jays form monogamous long-term bonds. They remain together all year round. They typically nest in a conifer, and both members help build the nest. The nest is a cup made of twigs, weeds, moss, and leaves held together with mud.

Red-Tailed Hawk: They are found all over the continent. Red-tailed hawks are known for their bright-colored tails. They prefer open areas such as fields or deserts with high perching places nearby from which they can watch for prey. They often perch on telephone poles and take advantage of the open spaces along the roadside to spot and seize mice, ground squirrels, rabbits, reptiles, or other prey.

Bar-Tailed Godwit: Another bird has made news (Zimmer, 2010). The bar-tailed godwit travels over seven thousand miles nonstop in nine days. The birds fly through the night slowly starving themselves as they travel 40 miles an hour. It's a surprising journey! It was reported in the *New York Times* Science section, May 25, 2010.

Lesson 13: Forecasting Weather

Topic of lesson: Predicting weather forecasts by blogging.

What do you want students to learn? Students should understand weather patterns, learn about weather instruments, study weather fronts, read weather maps, and create a weather forecast.

Why are the concepts important? Students learn how weather instruments help predict the weather. They study air pressure and learn about weather fronts. This information helps them become aware of the weather in their area. Review weather instruments such as thermometers and rain gauges and their functions. Introduce a barometer, which measures air pressure, and talk about how air pressure affects the weather.

What background information do students need before starting? Students need to be able to read a weather map. Students have studied clouds and the weather associated with these clouds. To provide a context for a lesson on weather fronts, students can observe various weather maps; the Weather Channel can help. Local newspapers often have a weather page tracking fronts for the day and week.

(For more information about weather fronts go to http://www .videopediaworld.com/video/34002/Atmosphere-Weather-Fronts).

Organization and procedures:

1. For each map, students record what they learned from the map and any questions they have. Students identify the different fronts throughout the United States in a given day. They check for the type of weather associated with the front.
2. Students make a list of questions about fronts. How does a front occur? What is the difference between a warm and a cold front.

What clouds are found in each? Students form groups to answer these questions. They become experts on their question. Each group shares its information so that the entire class has an understanding of weather fronts.

3. Students create a weather forecast predicting the weather for their area. They create a blog for their predictions. (For more information, go to edublogs.org.)

4. Students post their blogs and get reflections from other students.

Extension: Fellow teachers from other grade levels can be encouraged to write a class response to the weather forecasts they heard.

Connecting to the standards:

Science as Inquiry—ability to do science inquiry, understanding about inquiry

Earth and Space Science—changes in earth and sky

Science and Technology—understanding science and technology (NRC, 1996).

ORGANIZING THE SCIENCE CLASSROOM FOR YOUNGER STUDENTS

Water and sand center. Concepts such as volume conservation are easier to understand when children can measure with water and sand. Buoyancy can be explored with boats and with sinking and floating objects.

Blocks. Blocks are an excellent way to introduce children to gravity, friction, and simple machines. Leverage and efficiency can be reinforced with woodworking.

Painting. Finger painting helps children learn to perceive with their fingertips and demonstrates the concept of color diffusion as they clean their hands. Shapes can be recognized by painting with fruit and familiar objects.

Books. Many books introduce scientific concepts while telling a story. Books with pictures give views of unfamiliar things as well as an opportunity to explore detail and to infer and discuss.

Music and rhythmic activities. These let children experience the movement of air against their bodies. Air resistance can also be demonstrated by dancing with a scarf.

Manipulative center. Children's natural capacities for inquiry can be seen when they observe, group, sort, and order objects during periods of play. Children can pair objects, such as animals, and use fundamental skills, such as one-to-one correspondence.

Playground. The playground can provide an opportunity to predict weather, practice balancing, and experience friction. Children's natural curiosity will lead them to many new ideas and explorations.

Creative play. Dressing, moving, and eating like an animal will provide children with opportunities for expressing themselves. Drama and poetry naturally integrate learning about living things.

Literacy. The concrete world of science integrates especially well with reading and writing. Basic words, object guessing, experience charts, writing stories, and working with tactile sensations can all encourage early literacy development—while helping children develop an understanding of the natural world.

ENGINEERING LESSONS FOR KINDERGARTEN AND FIRST GRADE

At any age, being actively involved in the process of actually *doing* science moves learners along the road to scientific awareness. Scientific literacy should begin in the early grades, when students are naturally curious and eager to explore.

Engineering can reinforce science and math concepts—while being used to teach innovative thinking skills. For example, young children can read about the three little pigs and try to build and test "wolf-proof" houses. Index cards, cardboard, construction paper, cardboard, wood sticks, and paper cups all make good construction material. (The teacher can add or subtract items to make it more or less challenging.)

Each small group builds a house. The next step is huffing and puffing—trying to blow the house down. Someone with a lot of hot air can be the tester or a fan can be used at different speeds. Creative drama, movement, painting, writing a story or a poem are just some of the possible activities that can be used to express students' efforts and understandings.

SUGGESTIONS FOR DIFFERENTIATING SCIENCE INSTRUCTION

The teacher differentiates science instruction by attending to students' interests, learning styles, prior needs, and personal comfort zones.

Building a Ramp

In this activity, students are to create and test ramps using a variety of objects. Each small group has a chance to predict which slide will reach the barrier. Students will record their test trials and reflect on their results. The teacher can modify the problem and differentiate according to individual needs. One student couldn't get her ramp to work so she asked the student next to her for assistance in fixing it. It is important that students do not wait around, waving their hands to get their teacher's attention (unless the students nearby can't answer their question). At the end of the activity, students share reflections, problems, and successes.

Teachers can use tiered activities so that all students focus on basic understandings and skills—while at the same time working at different levels of complexity. By offering students access at varying degrees of difficulty, it is possible to get more out of each student; this way, each student comes away with essential skills and is appropriately challenged.

HINTS FOR HELPING TEACHERS DIFFERENTIATE LEARNING

It is one thing to agree with some of the ideas that surround differentiated instruction; it is quite another to have the approaches and strategies needed for making it happen. Here are a few ideas teachers can use to enhance instruction:

Evaluate students. The role of assessment is to foster worthwhile learning for all students. Performance assessments like portfolios work. So do informal assessment tools like checklists and anecdotal records. Also, teachers may use a compacting strategy to assess students before beginning a unit of study or teaching specific skills.

Consider the concept of difficult instructional tasks. Such tasks frequently involve

- free thinking and brainstorming
- finding what is essentially interesting to students
- acknowledging that there are many possible answers to problems
- using multiple intelligences in various situations
- connecting to natural things that can be seen, heard, and touched

Video technology can enhance science lessons: The wide availability of video has the potential for making learning more accessible for students who struggle in traditional classroom situations.

Practice using differentiated materials and active resources: The basic idea is to engage students who are reading at different levels, paying close attention to student interests and various learning profiles. A lesson might include simple concrete constructions—as well as more complicated abstract activities.

Introduce and employ science notebooks: Science notebooks are an important part of learning. Students can record data, facts, ideas, and concepts they learn. Notebooks might also include questions, observations, and predictions supported by evidence, conclusions, and reflections. A notebook provides a window into students' thinking and offers support for all students.

Construct an ongoing assessment plan. Use a record-keeping system to monitor what students do. Some students have difficulty with many things, while others have fewer weaknesses, but all have areas of strength. It is best to emphasize strengths, while attending to critical areas where they are having trouble.

In the classroom, it is possible to pull the content, processes, and products together when a student is struggling. Move on when important ideas and skills are understood.

PREPARING SCIENTIFICALLY LITERATE CITIZENS

Teachers have found that organizing scientific inquiry around real-life problems—the kind that can elicit critical thinking and shared decision making—excites their students. Inquiry today involves curiosity, observation, posing questions, and actively seeking answers.

One of the important goals of science education is to prepare scientifically literate citizens. This requires teaching students how to make use of scientific knowledge and understanding how it connects to daily life and society. Scientific literacy also involves having a broad familiarity with today's scientific issues and the key concepts that underlie them.

When it comes to science lessons in the classroom, strategies include concrete physical experiences and opportunities for students to explore science in their lives. There is an emphasis on ideas and thinking skills. Often, instruction is sequenced from the concrete to the abstract; occasionally, it works the other way around. Either way, students are actively involved in the collaborative learning process in a way that helps them develop effective oral and written communication skills.

Frequent group activity sessions are provided where students are given many opportunities to question data, design and conduct real experiments, and carry their thinking beyond the class experience. Students raise questions that are appealing and familiar to them, using activities that improve reasoning and decision making. Collaborative learning has become the primary grouping strategy where learning is done as a cohesive group in which ideas and strengths are shared.

Science can be an exciting experience for students and teachers when it is taught as an active hands-on subject. Connecting with other disciplines can provide many opportunities for integration with other subjects. Teachers need subject matter knowledge that is broad and deep enough to work with second language learners and others who may have difficulty with their school work. This often requires improving language and broad-based literacy development possibilities to get at content. It may take some effort to gain insights into others' experiences and ways they may be encouraged.

To understand and use science today requires an awareness of how science incorporates language and technology domains and how it relates to our culture and our lives. Good science teachers are usually those who have built up their science knowledge base and developed a repertoire of current pedagogical techniques. By focusing on real investigations and participatory learning, teachers move students from the concrete to the abstract as they explore themes that connect science, math, and technology.

Although aptitude varies, students can learn science—and they should have the chance to become scientifically literate. The science standards emphasize the processes of science and give a great deal of attention to reasoning, logic, evidence, and constructing explanations of natural phenomena.

Most twenty-first-century science teaching methods include many participatory experiences and opportunities for students to explore how science impacts their lives. The emphasis on inquiry involves posing questions, making observations, reading, planning, investigating, experimenting, explaining, and communicating the results.

Students develop effective teamwork skills as they work together, pose questions, and critically examine data. This often means designing and conducting real experiments that carry their thinking beyond the classroom. As instruction becomes more connected to students' lives, enriching possibilities arise from inquiring about real-world concerns.

TECHNOLOGICAL PRODUCTS OF SCIENCE: DEBATE THE PROS AND CONS

Whether it's good, bad, or in between, technology has always had a significant influence on behavior. New forms of media have frequently been denounced as threats to users' brain power and moral fiber. It is the same with electronic media.

On one hand, search engines may lower our understanding by encouraging us to skim the surface of knowledge, rather than go deeply into a subject. On the other hand, new digital media caught on for a reason. The knowledge base is exploding and information technologies (like the Internet) help us collect, select, and manage the collective intellectual output. A good debate is bound to come from discussing this topic. Simply pair students up and have them research one or both sides of the dilemma. Bring the conclusions to the whole class for a discussion.

On the plus side: computer use may provide some neurological advantage. For example, Internet users showed greater brain activity than non-users, suggesting they were growing neural circuitry. Also,

Internet users may become more efficient at finding information. Even some videogames may help with visual activity.

Every advance of technology comes with fresh approaches and trade-offs. For example, the convenience and, relatively, low cost of storing journals, books, and pictures through digital storage are appealing. Also, digitization allows for search within and across texts.

The downside is that some digital data can degrade faster than printed text. And the machines that can present the information rapidly evolve out of existence. A *New Yorker* cover sums up the dangers: an alien sitting on a posthuman pile of trash is reading the only thing that works—*a book.* Scattered around him are computers, CDs, VCR tapes, floppy disks, and mounds of discarded electronic gadgets.

The human brain is malleable and constantly being shaped by experience. This is usually seen as useful, but plasticity also allows our minds to be reprogrammed by our digital gadgets. Is there a danger of the human mind becoming a servant of technological processes? For example, even when briefly forced offline, some of us crave the stimulation of our electronic gadgets. And today, many are so distracted by e-mail and the like that they have trouble being fully in the moment, bruising deep thought, creativity, and innovation.

Attention, memory, and just about every aspect of learning are affected by our electronic devices. Attention is a good example because it involves what you let into your consciousness, what you remember, and what you forget. Clearly, heavy technology use opens up the possibility of pulling student attention away from deep thought and important skill development.

Does the excessive use of computers, cell phones, and the Internet cause users to become more impulsive, impatient, and forgetful?

New media changed the human relationship to information and ideas. The written word changed a verbal tradition—when the printing press came along, it amplified the changes. Radio, movies, and television distracted the mind with passive pleasure. Computers and their digital associates have moved it to another level. The web, for example, has encouraged us to trade away sustained attention for a frantic superficiality.

Another negative: the very existence of the online world has made it harder for many students to fully engage with complex ideas or complicated textual material. Also, the ability to focus can be undermined by bursts of information from e-mail, instant messages, online chats, and other electronic interruptions. Clearly, juggling all these stimulants changes how people behave.

A suggestion: do a little inquiry into these issues and see if you can come up with a list; draw a line down the middle of a piece of paper and put the pros on one side and the cons on the other.

SCIENTIFIC INQUIRY AND UNDERSTANDING THE NATURAL WORLD

Science is an evidence-based body of knowledge that continually extends and refines human understanding of the natural world. An inquiry approach to differentiated science instruction gives teachers a model that connects to the core of the science standards.

As far as student competency in science is concerned, learners need to know how to interpret and use scientific explanations. This includes being able to evaluate and generate scientific explanations and evidence. Also, it is important for students to actually use the tools of science (like math and technology) to engage in the practices of science. Along the path to proficiency, teachers need to provide plenty of opportunities for students to explore, apply, interpret, and assess themselves.

Active and collaborative science learning is a good way to connect students with the past, present, and future possibilities of science. Science learning can be enhanced with technology through data collection, simulations, webquests, the use of interactive tutorials, and so on. Differentiated instruction can also help, especially when science lessons are connected to real-life situations.

Teachers play a central role by promoting persistence, creativity, and innovative behavior. They structure experiences and arrange different levels of complexity that are based on students' prior knowledge and current needs. We now know that even children in the primary grades can think both concretely and abstractly. Although they have a lot to learn, students of all ages have reasoning skills that are related to scientific thinking.

We live in a time when science and its technological products are an ever present part of civic and political life. Without the chemical, physical, and biological fundamentals of science, people will believe just about anything. So it is appropriate that today's science curriculum plays close attention to responsible citizenship and self-understanding. It is also important to fully engage students in a way that encourages them to use scientific knowledge to make wise decisions and solve difficult problems that are related to life and living.

Science is increasingly interdisciplinary, with new fields emerging—biochemistry, biophysics, plant engineering, terrestrial biology, neurobiology, among others. More than ever, science has dimensions that extend to the social sciences—as well as ethics, values, and law. *Active observation*, *activating prior knowledge*, and *hypotheses testing* work as well across the curriculum as with scientific inquiry.

A moral imperative: making science instruction meaningful for all students in research-supported ways.

CONCLUSION: SCIENCE, CHANGE, AND THE FUTURE

Like other subject matter standards, the science standards call for a commitment to *all* learners. Teaching the full spectrum of students to think more deeply, solve problems, and acquire scientific skills and knowledge is a relatively new educational challenge. Experienced teachers know that paying attention to the varying interests and needs of students certainly helps. They also know that exploring the relationships between science and individual life experiences adds motivation to lessons.

When it comes to learning math or science, the focus should be on the process, not the answer. To make sure that students are really engaged in the practice of science requires teachers who can apply a wide range of pedagogical skills. The power of knowledgeable teachers is the key to success. But before teachers can invent the science classrooms of the future, they have to build up a reservoir of informal ideas, approaches, and techniques.

Throughout their professional careers, teachers need to be thoughtful, reflective, self-motivated, and lifelong learners. At any age, learning requires more than being able to answer questions; it requires knowing what questions are worth asking. It also helps if both teachers and students consider how science and technological innovation influence human welfare and economic change.

How can someone influence the future? One thing that can be done is to examine how change occurs. That way, you can at least contribute to reversing negative trends and minimizing catastrophic events.

There are certain similarities between our personal lives and social trends. We all have at least some control over the general directions of our lives and (to a lesser extent) over social trends. But it is someplace between difficult and impossible when it comes to foreseeing where our decisions and choices will lead us. Most of us end up someplace quite different from what was anticipated.

In some ways, change is a little like the historical record—the future will unfold a little like the wait, wait, and hurry up changes in life on this planet. For example, it took less than a third of the entire evolutionary span (the Cambrian explosion) to form the basic structure of most of the highly diversified life-forms that we know today.

There are many known and unknown components along the path to tomorrow and beyond that, even small interactions can alter the whole dynamic. Also, the element of chance plays a bigger role than most people realize. So, whether it's human or planetary history, there are such a vast number of variables involved that either gradual progress or slow reversals can quickly deepen or be reversed abruptly.

You can be sure that along with the gradual unfolding of long-lasting trends, there will be some sudden changes. But the only thing that is certain about the nature of the future is that it is impossible to make an accurate forecast.

REFERENCES

National Research Council (NRC). (1996, 2010). *National Science Education Standards*. Washington, DC: National Academies Press.

———. (2000). *Inquiry and the National Science Education Standards*. Washington, DC: National Academies Press.

———. (2007). *Taking Science to School: Learning and Teaching Science in Grades K–8*. Washington, DC: National Academies Press.

Tomlinson, C., Brimijoin, K., & Narvaez, L. (2008). *The Differentiated School: Making Revolutionary Changes in Teaching and Learning.* Alexandria, VA: ASCD.

Zimmer, C. (2010, May 25). 7,000 miles nonstop, and no pretzels. *NYTimes.com*, Science section.

RESOURCES

Abell, S. K., & Volkmann, M. J. (2006). *Seamless Assessment in Science: A Guide for Elementary and Middle School Teachers.* Portsmouth, NH: Heinemann.

Adams, D., & Hamm, M. (1998). *Collaborative Inquiry in Math, Science, and Technology.* Portsmouth, NH: Heinemann.

Altieri, J. L. (2010). *Literacy + Math = Creative Connections in the Elementary Classroom.* Newark, DE: International Reading Association.

Arthur, W. B. (2009). *The Nature of Technology: What It Is and How It Evolves.* New York: Free Press.

Athans, S. K., & Devine, D. A. (2010). *Fun-Tastic Activities for Differentiating Comprehension Instruction Grades 2–6.* Newark, DE: International Reading Association.

Blake, Jr., R. W., Frederick, J. A., Haines, S., & Lee, S. C. (2009). *Inside-Out: Environmental Science in the Classroom and the Field, Grades 3–8.* Arlington, VA: National Science Teachers Association (NSTA) Press.

Bloom, J. W. (2010). *The Really Useful Elementary Science Book.* New York: Routledge.

Buxton, C. A., & Provenzo, E. F. (Eds.). (2010). *Teaching Science in Elementary and Middle School: A Cognitive and Cultural Approach.* Thousand Oaks, CA: Sage Publications.

Callard-Szulgit, R., & Szulgit, G. K. (2006). *Mind-Bending Math & Science Activities for Gifted Students.* Lanham, MD: Rowman & Littlefield Education.

Carr, N. (2010). *The Shallows: What the Internet Is Doing to Our Brains.* New York: W.W. Norton.

Charlesworth, R., & Lind, K. K. (2010). *Math & Science for Young Children.* Belmont, CA: Wadsworth-Cengage Learning.

Davies, D. (2011). *Teaching Science Creatively.* New York: Routledge.

Doris, E. (2010). *Doing What Scientists Do: Children Learn to Investigate Their World.* Portsmouth, NH: Heinemann.

Eichinger, J. (2009). *Activities Linking Science with Math 5–8.* Arlington, VA: NSTA Press.

Evans, K., & Frazier, W. (2010). Blogging about the weather. *Science and Children* 47(8): 24–28.

Howard, L. (2011). *Five Easy Steps to a Balanced Science Program for Upper Elementary and Middle School Grades.* Englewood, CO: Lead and Learn Press.

Keeley, P., Eberle, F., & Dorsey, C. (2008). *Uncovering Student Ideas in Science: Another 25 Formative Assessment Probes.* Arlington, VA: NSTA Press.

Kim, M., Bland, L., & Chandler, K. (2009). Reinventing the wheel. *Science and Children* 47(3): 40–43.

Kovalik, S. J., & Olsen, K. D. (2010). *Kid's Eye View of Science: A Conceptual, Integrated Approach to Teaching Science, K6.* Thousand Oaks, CA: Corwin Press.

Llewellyn, D. J. (2010). *Differentiated Science Inquiry.* Thousand Oaks, CA: Corwin Press.

National Science Teachers Association (NSTA). (2006). *Linking Science & Literacy in the K–8 Classroom.* Arlington, VA: NSTA Press.

Norton-Meier, L., Hand, B., Hockenberry, L., & Wise, K. (2008). *Questions, Claims, and Evidence.* Portsmouth, NH: Heinemann.

Romberg, T. A., Carpenter, T. P., & Dremock, F. (2005). *Understanding Mathematics and Science Matters.* Mahwah, NJ: Lawrence Erlbaum Associates.

Rosenblatt, L. (2010). *Rethinking the Way We Teach Science: The Interplay of Content, Pedagogy, and the Nature of Science* (Teaching and Learning in Science Series). New York: Routledge.

Seeley, C. L. (2009). *Faster Isn't Smarter: Messages about Math, Teaching, and Learning in the 21st Century.* Sausalito, CA: Math Solutions.

Smil, V. (2008). *Global Catastrophes and Trends: The Next Fifty Years.* Cambridge, MA: MIT Press.

Starko, A. J. (2010). *Creativity in the Classroom: Schools of Curious Delight.* New York: Routledge.

CHAPTER 5

Technology and Education
Innovation, Digital Tools, and Preparing for the Future

Technology may be defined as the application of mathematical and scientific knowledge in a way that is designed to achieve some human purpose.

The historical timeline for technological tools can be slow moving, erratic, or even stationary. But there are many examples of sudden acceleration fueled by unexpected situations and surprising new breakthroughs. Take the Internet as an example; it was around for decades, but it didn't reach into popular culture until the World Wide Web came along in the early 1990s.

In preparing for tomorrow it is important to remember that human history is not necessarily linear or cyclical. More than ever, we have to prepare for the unexpected.

Imaginative new gadgets are one thing; figuring out how to turn them into productive tools is quite another. To put tech-tools to really good use there must be some idea of what kind of future we want to create. As far as educators are concerned, as much time has to be spent examining what education is as talking about how it should be delivered.

Math, science, and the human imagination have always been key ingredients in the invention and educational application of new technologies. But without a thoughtful human element, digital tools are likely to do little more than amplify unimproved outcomes.

When it comes to innovation, it's not just about whizbang gadgets. It has more to do with coming up with solutions to everyday problems you didn't know you had. The generation of imaginative new ideas frequently requires individuals and groups who are willing to push to the brink of failure to make something happen. It is hard to get every-thing the first time, so failure can be viewed as a type of practice. What is most important is learning from what doesn't work out and getting back on track.

As far as teachers and students are concerned, successful innovation requires vision, teamwork, time, and resources. It's a fantasy to as-sume that everybody must be prepared for a four year college. Even if such a miracle occurred, we would have to more than double the size of higher education to make room for the entire population. That is not going to happen. The important thing is making sure that all students have the academic and financial resources to be successful on the path they choose.

In today's hyperintegrated world many aspects of production can be accessed from anywhere by anyone. The main thing that cannot be outsourced is imagination. For a nation to do well in the twenty-first century, it must have an educational system that can nurture the ability to imagine and innovate. Just remember, there is a natural tension be-tween innovation (which by its very nature is disruptive) and traditional social arrangements.

When innovative individuals, companies, or nations encounter new problems the situation requires more than simply joining in the con-versation. A more imaginative goal is needed: finding transforming solutions and leading in new directions.

PROMOTING INNOVATION IN THE CLASSROOMS

In some classrooms today—and many more tomorrow—the intelligent use of digital technologies plays a key role in helping students ask the right questions. Among other things, these devices are constantly changing the way media is constructed and consumed. And they are bound to influence how new ideas are created and dealt with on the unstable frontiers of the twenty-first century. Randomness and uncer-tainty may rule, but in any conceivable future, innovative thinking and resiliency are bound to be important. You can also be sure that learning to think and thinking to learn will continue to be natural partners across the curriculum.

Teachers remain the key to a quality educational system. If they are well educated, intellectually curious, and fully aware of the characteristics of effective instruction, then students are much more likely to get a good education. Several years with good teachers will result in above average achievement—and several years of poor teachers are likely to put learners below their grade level. Of course there are many factors at play, including the academic background and motivation of students—to say nothing of parents, administrators, the media, and the community.

In spite of a host of influences, our focus is on K–8 curriculum and instruction. It takes skillful teachers to guide students through exciting learning adventures and discoveries, while at the same time learning a great deal from them. Imaginative activities for some and remedial drills for others are not an acceptable approach in today's schools. When it comes to encouraging students to generate imaginative new ideas, it is important that all students have the chance to engage in higher-level thinking, collaborative inquiry, problem solving, and meaningful communication.

Innovative approaches to problems or situations require a deep understanding of what is available today before it becomes possible to invent beyond them. Necessity can be a motivating factor. At other times, innovation involves a knack for navigating the space between desire and need. (Apple and Google are but two examples of tech companies that like to skate on that boundary.) But whether adults or children are involved, innovative behavior frequently involves discarding assumptions, building on the ideas of others, and tapping into networks.

You can be sure that the world of tomorrow will be a place where schools, the workplace, and just about everything else will be constantly shaking and sifting. In such an environment, a broad education has certain advantages over narrow vocational skill training because the former provides students with a perspective that helps them deal with an uncertain age. Clearly, the most innovative things often happen at the intersection of the liberal arts and technology.

NEW CAPACITIES FOR THINKING AND LEARNING

The minimum social price of admission to today's global economy can be paid only by people who are skilled in creative thinking, problem solving, and dealing with the unpredictable advance of technology. But no matter how reluctant you are, imaginative new technologies are an increasingly powerful and unpredictable force for change in the twenty-first century. So it is little wonder that understanding the implications of a world filled with the technological by-products of mathematics and science is coming to be viewed as a necessity.

Most of the jobs that today's students eventually take somehow involve computing and technology-related skills. Installing, upgrading, and using high-tech products in sectors like health care, education, and energy will be common. Pure technology jobs, like designing software and hardware products, will not show the same growth. But no matter what the shape of the future school or workplace, vision, teamwork, and resources are bound to be part of the innovative foundation.

A constantly changing set of digital technologies is providing new capacities for thinking, learning, working, and entertainment. These technologies are also dramatically changing the face of instruction and the very nature of childhood itself. So it is important to inject a little healthy skepticism into technology education and the use of these tools in other subjects. The healthiest approach: pay attention to the myths and the problems—as well as the magic.

A recent study by the Kaiser Family Foundation found that the average middle school student spends over seven hours a day with a computer, smartphone, or television. At home, Black and Hispanic children were found to be spending more time with electronic media (especially television) than White children. Also, some of the increase in media use over the last ten years has come from the use of mobile media. But whether it's mobile or stationary, heavy users tend to get lower grades (KFF, 2010).

Education is a team effort: parents, for example, have to shut down heavy media use at home. Teachers have to make sure that digital technology doesn't interfere with instruction at school. And at least to some degree, students have to take responsibility for their own learning.

The attention span of students may wither, but whether it's good, bad, or in between, electronic media can't be ignored. On the positive side, high-tech tools can be helpful in making subject matter more engaging and relevant for all learners. Also, there is at least the possibility of encouraging deeper thinking and problem-solving skills.

Now most communication devices are being built on a digital foundation. Like just about everything else, the usefulness of any of these devices at school depends on teachers who know how to avoid tech-distraction so that students can experience the advantages that technology can bring.

MATH, SCIENCE, AND TECHNOLOGY STANDARDS

The math, the science, and the technology standards all view technology as a means to form connections between natural and man-made worlds. There is also general agreement that it is important to pay attention to technological design and how technology can help students understand the big ideas of what is being studied.

The standards suggest that even in the elementary grades, students should be given opportunities to use all kinds of low-tech and high-tech technologies to explore and design solutions to problems. A suggested theme that reaches across subjects is helping students see the human factor and its societal implications. The laws of the physical and biological universe are viewed as important to understanding how technological objects and systems work. The same could be said for connecting students to the various elements of our technologically intensive world so that they can construct models and imaginatively solve problems.

Many teachers are familiar with the math and science standards, but fewer know about the standards that relate directly to information and communication technology. The International Technology Education Association's (ITEA) Standards for Technological Literacy suggest that within a sound educational setting, digital technology can enable students to become

- Capable information technology users.
- Information seekers, analyzers, and evaluators.
- Problem solvers and decision makers.
- Creative and effective users of productivity tools.
- Communicators, collaborators, publishers, and producers.
- Informed, responsible, and contributing citizens. (ITEA, 2000)

Educators know that an essential skill set in the twenty-first century is the ability to understand and work with technology. This doesn't mean electronic worksheets because they are just as boring as the paper variety. So it's important to make sure that our digital tools take us someplace more interesting than what's on traditional worksheets.

The research suggests that frequent and intelligent use of some digital devices can help students achieve important technological capabilities (Adams, 2001). Along the way, it helps if learners can develop the ability to sort through and evaluate the glut of information in cyberspace. It is also important to learn how to use data from search results to question, inquire, and solve problems. Like most things at school, making all these computer-related things happen has a lot to do with the capabilities of the classroom teacher. [International Society for Technology in Education (ISTE) NETS•T project.]

TECHNOLOGY AND THE MATHEMATICS STANDARDS

The National Council of Teachers of Mathematics (NCTM) Standards include the use of technology in their core suggestions about how to teach mathematics. This is not all new; calculators have long been part of most mathematics programs.

Calculators are recommended for school mathematics programs to help develop the following: (1) number sense; (2) skills in problem solving, mental computation, and estimation; and (3) the ability to see patterns, perform operations, and use graphics.

Calculators and other forms of technology continue to be used extensively in the home and office. The cost of calculators and other forms of technology continues to decrease, while their power and functions continue to increase. Curriculum documents increasingly encourage the use of calculators and other forms of technology. Some tests currently available allow and even encourage calculator use.

Exploring Math Activities Using Calculators

You may not be able to afford a computer for every two students. But for a tiny fraction of that cost, you can still get some interesting points across with cheap calculators. The new mathematics recommendations specify that calculators should be continually made available for all students. This includes homework, class assignments, and tests. The activity sets presented later in this chapter include some suggestions for how to use calculators and computers in your class. (These are examples of a design activity that meets the math/science/technology standards: standards 1, 2, 3, 4, 5, and 6.)

NEW TECHNOLOGIES AND THE SCIENCE CLASSROOM

Science is more than discovery, making things, and technological inventiveness. Scientists aren't successful without a tremendous amount of creative and innovative reasoning, because it's often not even clear what questions to ask—to say nothing of how pieces and answers might come together. Great scientists, like Galileo and Darwin, didn't introduce something into the world that was predetermined. They came up with imaginative and controversial new ideas that shaped the way we view the world. Adaptations, parodies, and borrowing have long been part of the creative process. Advances in technology follow a similar path.

The idea that science is composed of a catalog of facts that are not contested is false. Scientists know that there are no final answers and they usually acknowledge fallibility. The basic idea is to persuade, not compel. Science relies on the power of reason as new ideas are presented and tested in a rigorous process of peer review. Even established scientific "truths" are subject to change when new evidence and concepts come into play. In spite of the search for objective facts, we must all deal with the fact that for all their amazing power, math, science, and technology are profoundly human. Also, we must realize that when knowledge is created, it can open up a Pandora's box of good and bad possibilities.

If schools are to transform themselves, educators need to understand the innovative possibilities associated with the technological products of science. In addition, new skills are required for new media. Cell phones are but one example of the good and bad things that come along with digital devices. They can bring a great class to a screeching halt—or they can be used to gather information and view scientists at work. Cell phones used to be hardware driven, but now software and networking possibilities are moving in to control the scene. Yes, an array of digital technologies fuels today's growth segment, but don't bet on any one of them dominating the scene. It is more likely that lines dividing smartphones, computers, television, and the Internet will continue to blur.

With the recent improvements in wireless technologies, schools are doing more experimenting with all kinds of digital tools. Sometimes, it works out; sometimes it doesn't. A lot depends on knowledgeable teachers and the specific devices that students are trying to use. But no matter what the technology, educators now realize that online and offline learning sometimes require different skills. Work, on or off the Internet, can be done collaboratively. But one of the differences with more traditional lessons is that the Internet amplifies the need for students to be able to sift through multiple sources of information and figure out what sources are valid and useful.

Clearly dealing with a rapidly changing digital world requires new approaches and ideas. But how these new ideas get applied matters even more. How could the creativity and innovation success rates be doubled? The answer: fail more often.

Although our focus is more on the digital side of the equation, technology is more than computers. Technological tools that are common to mathematics and science include the graphing calculator, motion detectors, and scientific probes and should be included in the teacher's repertoire. CBL (calculator-based laboratory system) probes and graphing calculators capture real data and generate a scatter plot of data.

Good exploratory questions can be asked to generate more interesting functional relationships. You could, for example, ask students to create a linear descending line, an increasing line, a parabola, a horizontal line, and a vertical line using a motion detector. They will find this challenging, perhaps even impossible. But using probes and calculators allows students to look for patterns and to generalize many realistic formulas resulting from the graph of the data. The graphing calculator's statistical options allow for a formula or function relationship to emerge (Cathcart, Pothier, Vance, & Bezuk, 2007).

Although some of the lessons that follow later in the chapter can be accomplished without the use of technology, they can enrich the learning experience.

EXPERIENCING MATH, SCIENCE, AND TECHNOLOGY

Even when solving simple problems where they are trying to meet certain criteria, learners will find elements of math, science, and technology that can be powerful aids. At higher grade levels, lessons can include examples of technological achievements where math and science have played a part. Students can also be encouraged to examine where technical advances have contributed directly to scientific progress. To consider the other side of the coin, students can go online and find examples of where the technological products of math and science have done harmful things—like damaging the environment or taking away jobs.

Children and teens should have many experiences that involve math, science, and all kinds of technology. It might be as simple as measuring and weighing various objects on a balance scale. This can teach math- and science-related skills such as comparing, estimating, predicting, and recording data. What is the technology connection? A scale is one of the relatively simple technological tools used in mathematics and science for measuring mass or weight. Too frequently, however, teachers forget to mention the technological connection. Whether it's simple or complex, bathroom scales or hot new computers, technology is a ubiquitous part of our day-to-day world. And it's as misunderstood as it is hard to escape.

Students can be motivated by studying existing technology products to determine their function—identifying the problems the device might solve, the materials used in construction, and how well it does what it is supposed to do. A low-tech device, like a vegetable or cheese grater, is an example of a simple object that young children can investigate. They can figure out what it does, how it helps people, and what problems it might solve and cause. Such a problem provides an excellent opportunity for directing attention to a specific, yet simple technology. In the early elementary grades, many tasks can be designed around the familiar items in the home, school, and community. In the early grades, problems should be clear and have only one or two solutions that do not require a great deal of preparation time or complicated assembly.

As the standards documents make clear, children can learn a great deal about math, science, and technology from the low-tech and the high-tech ends of the spectrum.

Many curriculum programs and some state guidelines suggest that teachers integrate math and science with digital technology and examine related social issues. The suggestion here is that this should be done in a way that encourages multidisciplinary analysis of problems that are relevant to the students' world. To get a little history into the process, it is also useful to examine many of the good (and bad) technology-assisted events of the past. A question: how many electronic communication tools available today weren't around ten years ago? Thirty years ago? Point out that technology is in a constant process of adjustment, innovation, and surprise.

ENCOURAGING PROBLEM SOLVING AND INQUIRY

A sequence of five stages is a good way to approach the problem-solving process: (1) identifying and stating the problem, (2) designing an approach to solving the problem, (3) arriving at and implementing a solution, (4) evaluating results, and (5) communicating the problem, design, and solution. In keeping with the math, science, and technology standards, teachers are increasingly encouraging students to design problems and conduct technological investigations in ways that incorporate interesting issues and dilemmas.

In a world permeated by technology, students will do better in school and in life if they are familiar with technology and have an understanding of the subject. Those who graduate with a better technology education are likely to improve their choices as consumers, employ technology more effectively in their daily lives, and make better citizen choices about technological issues in society (NRC, 2000, 2001). In 2000, the International Technology Education Association (ITEA) published *Standards for Technological Literacy: Content for the Study of Technology*.

AN OVERVIEW OF ITEA'S *STANDARDS FOR TECHNOLOGICAL LITERACY*

Theme 1: Nature of Technology
- characteristics and scope of technology
- core concepts of technology

- relationships among technologies and connections to other subjects

Theme 2: Technology and Society
- cultural, social, economic, and political effects of technology
- effects of technology on the environment
- role of society in the development and use of technology
- influence of technology on history

Theme 3: Design Processes
- attributes of design
- engineering design
- role of research, development, invention, innovation, and experimentation in problem solving

Theme 4: Abilities for a Technological World
- apply the design process
- use and maintain technological products and systems
- assess the impact of products and systems

Theme 5: The Designed World (knowledge in specific technological fields)
- medical
- agricultural and biotechnology
- information and communication
- transportation
- manufacturing
- construction

SCIENTIFIC REASONING AND THEORIES OF CHANGE

Enlightenment philosophers like Descartes, Rousseau, and Voltaire believed in shining the light of reason on superstition and fundamentalist religion. The basic idea was to do away with old precedents and come up with new ideas. Inspired by the scientific revolution, they put their faith in the ability of human reasoning to logically arrive at universal truth. A good example is the seventeenth-century mathematician René Descartes. He thought that it was best to discard accumulated prejudices and erect one logical certainty on top of another.

In England, Adam Smith, David Hume, and Edmund Burke took a more gradual step-by-step approach to change. This theory paid more attention to pointing out the limits of reason and examining sentiment and other natural desires. They also believed in a more gradual approach to changing what occurred previously, one step at a time.

A good discussion can be generated regarding the nature of change. Take synthetic biology as an example. Biologists can now use digital technology to manufacture a genome that exists nowhere in nature and insert it into a blank bacterial cell. The belief is that we are getting close to having the ability to design and make just about any sort of life-form that we want. What classical thoughts about change come into play here?

The basic idea of synthetic biology is that organisms can be broken into a set of parts. DNA comes in specific pieces (genes)—and genes contain instructions for making proteins and manipulating molecules. Like a computer, the parts come in various shapes and sizes, making it possible to design elements in a way that controls the functions of a cell.

Since they start with a known set of genes (a genome) that occurs in nature, synthetic biologists are assisted by more than three billion years of evolution. After the genome is put into a live cell, ingredients are added to power up the synthetic variant.

Like it or not, scientists now have both the knowledge and the tools to design creatures that do not exist in the wild. This may be helpful in rapidly coming up with a new flu vaccine or an organism that eats crude oil. But being able to write in a genetic language of our own design raises all kinds of ethical issues that need to be examined.

Discussion: How do we safely manage the synthetic biological constructs that we bring to life?

CONCENTRATING ON THE ESSENTIALS OF INSTRUCTION

Although the essentials of learning matter more than novelties of new media, the complete avoidance of digital technology is not an option. There is, after all, general agreement that the key to a nation's success lies in having an educational system that can nurture imaginative students. And these students have to know about technological tools and work well in collaborative contexts that are creative, innovative, and flexible.

Computers and their attendant technologies, at least, have the possibility of evolving in ways that help students acquire the ability to reason, empathize, and come up with new ideas. Digital tools can also open up possibilities for more active and differentiated engagement with the subject matter. Digital technology can help by allowing for collaboration with peers at home and around the world.

Whatever the educational media of choice, the rules of the instructional game remain the same. Teachers who understand the characteristics of effective instruction know that students learn more successfully through interactive, collaborative, student-centered learning. When it comes to dealing with innovative possibilities, children and teens need to develop skills that help them adapt to rapidly changing realities.

The brief history of digital technology is filled with people who made confident predictions about the best routes for the future, believed their predictions, and had trouble when things didn't work out quite the way they thought they would. Still, innovation requires approaching the future by thinking broadly, being bold, and appreciating the strategies for innovation that worked in the past.

As far as national policies are concerned, to remain relevant countries have to take the resources they have and build on proven educational models. In one sense, there is an increasing innovation gap between the world's richest and poorest societies. But if you look closely, the gap is more evident in the application area than it is in design or educational quality. As advanced economies of the world build ever-faster bandwidth networks and fancier digital devices, their counterparts in the undeveloped world are finding more uses for cheap computers and basic cell phones.

To Americans, complexity has often been associated with progress. So why are many complicated digital devices and complex educational programs being called into question? In many areas, complexity and complication grow a step at a time. Each step along the way may seem reasonable when it's taken. But when everything is added together, it becomes confusing and stifles innovation. Simplicity isn't necessarily the answer, but the best solutions in today's world are often those that are just complex enough to work.

BUILDING ON THE SOCIAL NATURE OF LEARNING

Many classroom teachers have found that Common Sense Media is a good source of information (for teachers and students) about the social issues involved. Technology educators also suggest including some of the larger structural issues found on the Internet. (Everything from web viruses to how companies profit from tracking consumers online falls into the "structural issues" category.)

Whether it is chat rooms, web threads, or face-to-face contact, you can get a good discussion going about whether Internet sites like Facebook, Myspace, and Twitter can empower a type of "group think" that shuts out the voices of adults, history, current events, and so on. By fourth grade, many students are experimenting with social media and the constant babble of thoughts found there.

Social networking sites help students discover new "friends" by discovering common interests, skills, and geographical locations. Facebook and Myspace are two examples.

- *Instant Messaging* is a type of real-time communication between two or more people. It is typed text conveyed by computer over the Internet or from cell phone to cell phone.
- *YouTube* involves sharing videos with friends, family, and the world. It's free user-generated video.
- *Twitter* is a free social networking service that allows users to send "updates" or "tweets." These text-based messages can be up to 140 characters long and posted on the Twitter website. (It doesn't seem to be a favorite of teens—and adult users often don't stay with it all that long.) Now the Library of Congress is making a permanent record of millions of tweets. They may be boring when read one at a time, but perhaps, when they are read as a composite group (in the future), there may be more meaning in them (Kirkpatrick, 2010).

You can "friend" someone on Facebook, "favorite" a video on YouTube, and "follow" someone on Twitter. And you can just as easily get rid of them by adding *u* and *n* to the beginning of each word. "Unfriend," "unfavorite," and "unfollow" are now three verbs in the realm of social (web) interaction (Turkle, 2011).

Suggestions for Parents

Parents need to realize that children and teens have done some strange things with social networking sites like Facebook.

- Set up the rules of social networking early—before bad habits become ingrained.
- Keep computers out of the bedroom so you can glance at e-mails, pictures, and so on.
- Set limits on the use of digital gadgets on school nights. (Thirty minutes is a good limit.)
- Take the cell phone and other gadgets away for eight hours at night (no 3 a.m. texting).

The most important job for parents is to keep their youngsters safe— even if children become angry when they aren't allowed to keep a cell phone under the pillow.

Making smart use of digital devices is a shared responsibility. What makes us human is our social nature. The concern here is not whether social networks have the potential to make a personal or educational contribution. Rather, we worry about social networking sites hurting children and lowering the quality of intellectual discourse.

TWENTY-FIRST-CENTURY "TRANSPARENCY"

When the Internet took off in the 1990s, users took it for granted that it was wild, free, and anonymous. A popular *New Yorker* cartoon from that period sums it up with a cartoon of one dog telling another: "On the Internet, nobody knows you're a dog." Now it's harder to conceal identities; if you're a mutt, there are ways of figuring out your pedigree.

Twenty-first-century "transparency" has a lot to do with the way digital technology records more and more of our behavior for all the world to see. As everything flies up into the cyberspace "cloud," it becomes possible for anybody to hack your e-mail account and just about anything else. Even your location is available thanks to an array of mobile devices, ATM cards, tracking technologies, and Internet-based databases. The whole thing is creating a connected societal environment in which we are all under a constant threat of surveillance.

Young people need to know about the possibility of negative consequences if they post personal things online. Loss of privacy and predatory behavior are two of the many negative possibilities. Another thing that youngsters often miss is the fact that what they put online is semipermanent. They need to know that the line between private and public spaces gets blurred online; the result can be personally embarrassing today *and* tomorrow.

As students consider what to post online, ask them if they want their parents, friends, or teachers to see what they post—because sooner or later, they might. Remember, if something is online, it doesn't go away and it's easy for anyone to get access to it. A chat room, for example, is only as private as every member wants it to be. Using the Internet for personal connections is a little like crossing the street—you have to carefully look both ways first.

From the invention of the printing press to the telegraph, to radio and television, and to computers and the Internet, innovation has always generated both good and bad possibilities. On the negative side, the Internet makes it possible for people to feel fully informed without ever encountering an opinion that contradicts their prejudices. In many ways, the web is a vast echo chamber that allows fringe voices to get more attention than they deserve. Still, although extremist groups may exploit the web and governments may manipulate it, the global promise is limitless. Much depends on the preparation, perceptions, and action of its users.

Whether use is good, bad, or in between, everything from websites to tablet computers is bound to make a real difference on many levels. But digital technology can never deliver us from ignorance or evil. Only good people with a sense of direction and an understanding of both the media and the message can do that.

DIGITAL TOOLS IN TODAY'S CLASSROOM

Teachers and students may wish to communicate with peers, parents, and the larger world community. However, we suggest that teachers not accept instant messages from students; reserve your personal e-mail address for professional activities. Teachers can, however, use the full range of available technology to enhance their productivity and improve their professional practice. Lifelong learning and personal change are facts of life in the twenty-first century. When teachers are supported and well prepared, they can walk in the world with such confidence and enthusiasm that they don't have to fear the unhappiness of change.

New technology is constantly changing math, science, and technology instruction by altering the general academic environment and providing new opportunities for students to create knowledge for themselves. This encourages students to learn by doing—going beyond the "telling" model of instruction that so many students find problematic. Digital technology can also serve as a vehicle for inquiry-based classrooms—giving students access to data, experiences with simulations, and the possibility for creating models of fundamental math/science/ technology processes. Like a good teacher, today's technology has at least the possibility of increasing everybody's capacity to learn.

What about early childhood education? In the primary grades, technology has gone beyond simply being an enrichment experience or an add-on to the curriculum. Digital technology can help younger children view things in their world from different perspectives. With a digital camera or microscope attachment to the computer, learners can view and communicate close-ups of plant and animal life. Also, children in the lower grades can hypothesize, experiment, and work together to solve problems with a variety of technological tools. At any age, when students listen to the views of others, they stretch their concepts, consider different points of view, and recast ideas to communicate and advance their understanding.

Digital technology is transforming our educational and social environment so fast that we haven't had the time to think carefully about the best way to use it. Like everyone else, teachers are consumers of technology and they need to be able to judge critically the quality and usefulness of the electronic possibilities springing up around them.

Many people outside of school think that life today is moving too fast—hyped up with wireless laptops, netbooks, tablet computers, cell phones with TV shows, podcasts, blogs, BlackBerries, and instant messages. They should try to imagine what it is like to be a teacher with struggling students, new curriculum choices, political demands, standardized tests, and digital technologies swirling around them.

What students learn from critically and creatively examining any visually intensive medium will apply as more advanced multimedia devices come online. Whether it is television, the Internet, or just about any other platform, students' social, educational, and family context influences the messages they get, how they use media, and how "literate" they are as users. To become critical and imaginative thinkers who can understand conflicting media messages, students should be able to

- Understand the grammar and syntax as expressed in different presentation forms.
- Develop ways of looking at problems that focus more on context and less on reductive answers.
- Analyze the pervasive appeals of advertising.
- Compare similar presentations or those with similar purposes in different media.
- Identify values in language, characterization, conflict resolution, and sound/visual images.
- Identify elements in dramatic presentations associated with the concepts of plot, storyline, theme, characterization, motivation, program formats, and production values.
- Utilize strategies for the management of time online and viewing and program choices.

Whether it's understanding media or contemplating the uncertain nature of a rapidly changing world, it's a little like getting into good physical condition—you have to begin very early. And you have to keep at it.

Influencing the settings in which children attend to media is a crucial factor in productive use. For example, turning the computer or TV set off during meals sets a family priority. Families can also seek a more open and equal approach to choosing Internet sites or television shows—interacting before, during, and after the viewing. Parents can also organize formal or informal activities outside the house that provide alternatives to computer games, Internet use, or TV viewing.

It is increasingly clear that the education of children is a shared responsibility. Parents need connections with what's going on in the schools. But it is teachers who will be the ones called upon to make the educational connections intertwining varieties of print and visual media with science, mathematics, or technology education.

It is possible to explore any medium in a way that encourages students to become intelligent consumers. The activities included in this chapter are designed to be used with upper elementary and middle school students.

Questions to Start the Discussion

1. What is your favorite website, television show, and movie?
2. What kind of information, show, and movie is it?
3. What are the formal features of your choices?
4. What are the most appealing elements of each?
5. What do you know about how each medium constructs its "stories"?
6. What are some of the formal and informal structures of the Internet, movie industry, and television broadcasting?
7. What are the values in these mass-produced "programs" and how do they change our shared experiences as a people?

Fifteen years ago, our shared experiences almost always included books and newspapers. Except for short passages read and written on the Internet, only the more educated members of our society now spend a great deal of time with the printed word.

USING DIGITAL TOOLS TO EXPLORE QUESTIONS AND DILEMMAS

Making good use of technology requires going beyond using it as an electronic version of the same old thing. At its best, technology can deepen subject matter understanding and promote innovative thinking. Digital technology can also open up new and unique possibilities that take students well beyond textbooks and worksheets. Among other things, it can amplify the energy and the intellectual curiosity that is needed to learn more on a daily basis.

Technology opens up possibilities for moving schools in the direction of being more responsive, modern, and effective. There are also many ways that it can help teachers. For example, a webcam can be used to record a difficult lesson. The next step would be to put it up on the Internet and get some suggestions from other educators. Students can take a similar approach when they come up with difficult questions; remember, if they can't get answers, their curiosity is diminished the next time around. In addition, digital tools can help learners get around what's blocking answers—while at the same time increasing motivation.

For students to successfully sail through the crosscurrents of our transitional age, teachers need to make sure that they are given the opportunity for important learning. There are times when digital technology can help; however, at other times it may be best to get it out of the way. A lot depends on the nature of the digital device and the structure of the curriculum.

New mobile products are coming out all the time and polite use is hard for some to figure out. As irritating as those obsessed with handheld devices and computer generated e-mail may be, no one can completely dismiss the technology. Here are some rule suggestions for educators and young adults:

- Set times of the day when you check messages; at other times, put devices away.
- Text messages lead to more text messages; use the phone or show up in person.
- If you have a handheld device with you at a meeting, place it face down and off.
- Set up an e-mail filter that keeps out ads and low-priority messages.
- Avoid replying to messages on evenings or weekends.
- Make sure that instant answers are not expected.
- Be sure that *you* (not the device) are in control.

At any age, it's important to have some rules of the road so that using digital devices shows respect for those around you.

DIFFERENTIATION, THE INTERNET, AND INTERESTING GROUP WORK

Differentiation, collaborative learning, and the Internet are natural partners. To access a good sample of recommended Internet resources try www.filamentality.com and search for "DI using technology." But whether you are online or offline, it's important to remember that students come to whatever they are doing or reading with different levels of prior knowledge. To find good activities, we often use webquests. These are investigative activities on the Internet that are educator created and peer reviewed. For a few free examples go to www.webquest .org and www.discoveryeducation.com/teachers. If you are willing to pay for a subscription, www.webquestdirect.com.au is a good service.

Although you wouldn't know it from the way many of us write about the topic, the *Internet* and the *web* are not synonymous. The Internet is a worldwide network of computer networks, which we have been using in order to download information from many sources. The World Wide Web, on the other hand, is a system of interlinked hypertext documents, which can be viewed using a web browser such as Firefox or Internet Explorer.

Wired magazine's Chris Anderson and at least a few others believe that the Internet and the World Wide Web are splitting up (Anderson & Wolff, 2010). This is because of the recent rise of smaller and lighter mobile devices such as the iPod, the iPad, BlackBerries, and others that are increasingly becoming the appliance of choice for both personal and work activities. The World Wide Web is just one application that exists on the Internet and uses the Transmission Control Protocol/Internet Protocol (TCP/IP) as a means of moving information packets around. (A *protocol* in this case refers to the guidelines that allow a "digital handshake" among computers that enables the exchange of information.)

With the increasing use of the mobile devices, it has been reported that the use of the Internet browsers most people are so familiar with is being superseded by more popular applications or "apps" that no longer need a browser and enable things like e-mail, Skype calls (free Internet-enabled phone calls), online gaming, iChat conversations, Netflix movie streaming, and so on. It is also expected that the number of users of small portable devices will eventually surpass the number of users of personal computers and laptops.

A simple lesson for mixed-ability groups: everyone reads or does the same math, science, or technology problem, activity, or section of text. Each student finds a partner and does some Internet research that they will bring back to the small group. We sometimes have upper grade students explore the medium itself by going online to read reviews of related books.

The Google Book Search Library Project has made a vast number of books available online. Information from the *outside* world is readily available for students. It is even possible to "google" human genes to get information about the world *inside*. Disruptive technologies, like the iPad, are good topics for student exploration. They shake things up by changing the way that students, adults, and businesses think and operate.

Exploring the differences between successful technology companies provides insights into the world of technological innovation. Apple and Microsoft are good examples. When Apple falls short, it is often the result of paying more attention to aesthetics than to the practical needs of users. With Microsoft, when functionality falls short it often reflects a failure of engineers to empathize with the needs of users.

When it comes to encouraging innovative thoughtfulness in the classroom, it is important to provide multiple (differentiated) options for accessing information. Working with peers to engage content with digital tools is part of the process. The goal is to inspire learners to be thinkers and innovators. So it is important to go beyond chalk, teacher talk, and printed materials in ways that build on divergent thinking and group dynamics. Collaborative groups help strengthen the classroom community and do a better job of engaging the full spectrum of students. When students feel that "we're all together," it encourages them to take on different roles, share resources, and help each other learn.

THINKING, LEARNING, AND THE UNPREDICTABLE REACH OF TECHNOLOGY

When hearing the word *technology* many think only of computers and the Internet. A minority think of technology is the process by which humans modify nature to meet their needs and wants.

As far as the schools are concerned, technology is more than computers and spacecraft. The knowledge and processes used to create and operate core technological products are an important part of technology education. Also, students need to know that technology is more than the application of math and science. It may build on, complement, and support those subjects, but it has a wider reach.

Definitions of technology education and educational technology vary and concepts overlap, but there is general agreement that learning with (or about) technology is much more than computers. It involves exploring the technological knowledge and skills used in the designed world—transportation, medical technology, and transportation technology, and the like.

Technology education deals with the design processes and technological abilities that are applied to a wider range of human needs. Educational technology is a slightly different subject. It may be viewed as using technologies in the education of students: computer programs, the Internet, calculators, and even things like heat probes in science.

Since backgrounds and interests vary greatly, students need to approach technology in different ways to appreciate and understand the subject. This requires classroom practice that allows for the differences in student interests, prior knowledge, socialization needs, and learning styles. In the differentiated classroom, teachers take care to vary the degree of structure in a lesson—as well as the pacing, complexity, and level of technical abstraction. The basic idea is that learning is most effective when adjustments are made so that a learner at any achievement level can make sense (meaning) out of whatever is being taught.

If students are encouraged to whimsically gaze into the beyond, they just might get a little of it right. The process should be a search for the truth, not wishful thinking. Evidence must trump personal opinion. Just make sure that learners are flexible and resilient enough to deal with the surprises—especially when the data contradicts conviction. Even a failed project or experiment doesn't have to be a wasted effort—especially if something is learned from the failure.

Part of the excitement over what is just beyond the horizon is not knowing when the boundaries of effectiveness will shift and where things will end up. One example of a surprise that awaits us is associated with learning about the architecture of information storage in the human mind. Could self-aware digital systems eventually emerge from the interconnections of computers and computer networks? From problem solving in mathematics to understanding the natural world (science), technological designs and tools have constraints that limit our choices.

Most users of today's personal computers come nowhere near making use of their capacities. Our digital tools are becoming capable of dealing with more than our intellects put forward. It's just that we usually don't explore or push these gadgets to do new and more imaginative things.

Where technology and its associates are taking us remains a mystery. At least a few of the consequences can be predicted. Many cannot. For example, who at the beginning of the twentieth century would have predicted the human consequences of physics and the technologies associated with atomic energy? In another fifty years, the same thing may be said about posthuman evolution if machine intelligence reaches a stage where it can play a major role in technology innovation.

The twenty-first century is one of those times when we all have to look out for unexpected devices and unusual events showing up on the horizon. As the power of new hypermedia becomes increasingly intense—and politics ever more rabid, it's important not to wait around passively while unthinkable events overwhelm us. To paraphrase Thomas Edison, if we all did things we are capable of, we would literally astound ourselves.

TECHNOLOGY, STANDARDS, AND SOCIAL VALUES

The math standards and the science standards make it clear that both high-tech (computers) and low-tech (simple manipulatives) are essential to problem solving and inquiry. The technology standards focus on the digital side of the equation. All of the standards suggest ways for technology to improve instruction for all students. A major point is that every student should have access to high-quality content so common learning goals can be reached.

When it comes to math and science, the underlying assumption is that students should not be limited to executing basic rules and remembering fundamental concepts. Today's technology instruction follows the same pattern. Aiming low and slow just doesn't get the job done. When in doubt, it is best to "teach up" with strategies that engage the imagination with the help of active learning and group participation.

Education, like technology, both shapes and reflects the values found in society. Take the example of digital technology where learners can be isolated or encouraged to join with others. Those who think that it is all good or all bad miss the point.

In our personal and civic lives, technology can slip through our hands to limit our choices at work and erode the edges of the constitutional rights of privacy in our daily lives. It can bring out the worst in human nature and diminish the imagination. On the other hand, technology can empower individuals, encourage students' curiosity, and spark innovation.

Online and printed newspapers and other sources have plenty of single frame cartoons (often with political or social commentary). Just take the caption off and let students come up with their own or keep the caption and have them draw the cartoon. In spite of laugh lines, misplaced enthusiasm, and nuisances, technology is both a concern in its own right and an essential part of the math and science curriculum.

Like all technology through the ages, digital technology is a double-edged sword. At their out-of-school worst, games are rigidly preprogrammed arcade-like shoot-em-ups where children frantically click on icons for instant gratification. At their worst, computers and the Internet can turn mathematics, science, technology, or any other topic into a spectator sport. An hour of *PowerPoint-and-talk* isn't much better than an hour of *chalk-and-talk.*

At their in-school best, technologies can be an excellent vehicle for individual and small group questioning, investigating, analyzing, communicating, and simulating content-related situations. Technological tools provide all kinds of possibilities for students to take control, solve problems, inquire collaboratively, and observe phenomena that would otherwise remain unobservable.

INNOVATIVE DIRECTIONS WITH
TECHNOLOGY AND ITS ASSOCIATES

To get the educational job done requires teachers who can help students from diverse backgrounds gain the competencies needed for identifying, analyzing, and solving mathematical, scientific, and technological problems. With all of the high-tech explosion of possibilities, it is important to realize that curriculum connections to the world of numbers and nature must be filtered through the mind of the teacher.

It is important to make sure that students who struggle with math, science, or technology become engaged with these subjects. Clearly, investment in "human ware" beats investment in "software" every time. Many countries, including this one, are always just one generation away from a dismal national decline. To avoid an education-influenced disaster, it is important to focus on what is wrong and *pay special attention to the successes.*

The math, science, and technology standards connect students to the designed world and introduce them to the laws of nature through their understandings of how technological objects and systems work. People have always invented tools to help them solve problems and find answers to the many questions they have about their world. Just as scientists and engineers work in teams to get results, so students too should work in teams that combine math, science, and engineering talents.

The science standards are a good example. They begin with the suggestion that students should learn to understand the design process and be able to solve simple design problems. The process is helped along if students develop the skills needed to creatively solve everyday problems with technology; it also helps if they are familiar with technological products and systems.

One way to foster innovation and creativity is to encourage learners to take two unrelated ideas or things and come up with new solutions, objects, or situations. To do this it helps if you can look at the world as multiple objects mixed in multiple ways. Creative professionals in math, science, technology, and other subjects often go from looking at one thing from multiple perspectives to viewing multiple things mixed together in unpredictable ways.

It is usually best if students of all ages learn about math, science, and technology by firsthand experiences with technological tools similar to those used by accomplished experts. Although we pay close attention to digital technology, the math and science standards point out that young children should also see the technological products and systems found in the relatively low-tech world of zippers, can openers, and math manipulatives. Throughout the primary grades, students can use both low- and high-tech approaches to engage in projects that are appropriately challenging for them.

QUESTIONS, POSSIBILITIES, AND PROBLEMS IN CYBERSPACE

Small media, like YouTube and Facebook, are having a major effect on popular culture. In a digital era of new cognitive models, social networking websites, and freewheeling video downloads, one person's voice can occasionally have as much power as a TV network. When a new medium arrives on the scene there is usually the same reaction: What is it going to take the place of? What is it going to do to culture and morals? Is it going to make us sick? Some of the pluses and minuses of digital technologies may be found in earlier media. For example, when television first gained a central place in the American consciousness, the sociologist Leo Bogart wrote that it was a "neutral instrument in human hands. It is and does what people want." The same thing might be said about today's multimedia and telecommunications technologies.

As Google goes about digitizing the world, it has to deal with the tensions that arise from both age and cultural differences. Students can learn a lot from exploring the different national approaches to the issue of privacy. In Europe, for example, privacy is viewed as a human-dignity right. The basic idea is to try and keep the mass media out of people's lives. In the United States, privacy often has more to do with keeping the government from overreaching. And sometimes, it is enforced as a consumer protection issue. In China, the government tries to keep Internet users away from pornography, violence, and certain religious/political sites.

Whatever culture you are in, technology has changed the way personally identifiable information is made available. Even if you don't disclose personal information, friends and others online will do it for you. Whether it is by accident or on purpose, everything from references to your school and employer to your sex and interests gets out into cyberspace. Since changes in digital technology make life increasingly transparent, students need to be made aware of the fact that patterns of online social communication are revealing.

Technological change is more than a random phenomenon; usually there is a cause and effect relationship. Rapid changes in information technology are resulting in less of a difference among the television screen, the computer screen, and cell-phone-linked networks. What we're seeing is a steady stream of fast-moving crowd-sourced innovation, and collective creativity.

LEARNING ABOUT TECHNOLOGICAL ADVANCES

Learning with (and about) technology has a lot to do with past experiences and whether or not a student's experiential background generates possibilities for innovative behavior. At home and at school, much of what an individual learns is shaped by social interaction—including shared webs of ideas, symbols, and actions. Innovation depends on the skill of the individual and the skills of the group.

Continuous interaction and flexible experimentation are big difference makers. But it's not just the quality of ideas, teamwork, or new technology. Successful innovation also has a lot to do with how all these factors interact with the world.

Throughout history, changes in media have affected our sense of time and space. The computer, the Internet, and their accessories are more powerful than the usual pace of history; they represent a great leap forward in communication and information technology. As virtual communication pushes actual human interaction into the background, we spend more time apart. For children and young adults, the social rhythm of learning is often disrupted.

Each new technology has its pluses and minuses, but in a digital world, natural speed limits are often exceeded. Our students' view of the world is shaped by high-tech factors like e-readers, smartphones, and tablet computers. At other times, technology can support technological transitions—much like Amazon's Kindle encouraged others to develop e-readers and Apple's iPad encouraged the move to tablet computers.

Change is coming on faster than ever—and it is merciless with those who aren't prepared and flexible. So it is important to get ready for new complexities, fresh interactions, and the speed of paradigm-shifting events when they do happen. Part of getting prepared is identifying needs, asking the right questions, testing new ideas, and being aware of what's on the horizon.

When it comes to helping students develop an imaginative under-standing of technology, the key is the content of technology-related lessons and how they are connected to what is going on in the classroom and the world. A major instructional goal is making sure that students can apply computer and technology skills to a wide range of problems and situations. Technology classes themselves often spend too much time on word processing and spreadsheet software. We suggest paying more attention to the magic of computing and how digital tools reach across academic subjects, industry, and society.

ACTIVITY SETS FOR MATH, SCIENCE, AND TECHNOLOGY

A Teaching Technique: Scaffolding

When students falter in a lesson, a teacher can help just enough to make sure they accomplish a task. Later, the learners will be able to do the work on their own.

Steps:

- Observe to figure out what the student or pair of students is trying to do.
- Ask the student(s) what they intend to do. (Getting verbal input shows respect for ideas.)
- Comment on the student actions in a way that shows you are paying attention.

- Scaffold verbally or through action to provide possible ways to solve a problem.
- Once the process of completing the task is under way, step back from input.
- When work is done, comment on the accomplishment. (Your comment will add value to the new skills.)

The structure can be differentiated in a way that supports what's needed to accomplish the task. The number of steps needed to solve the problem can be reduced (or increased) so that interest is maintained and frustration is avoided. Peers can be very helpful in the scaffolding process. The teacher focuses on what the student is doing, what he or she needs in order to learn, and only gives enough help to overcome roadblocks.

Set 1: Math Activities

Using the Team Approach with Math Activities

1. Use calculators to improve basic estimation skills. Form teams of students. Be sure each student has a calculator. One member from the first team says a three-digit number. A player from team two selects another three-digit number. Both players write an estimate of the sum of the two numbers. Students are limited to responding within five seconds. Both players then use the calculator to determine the sum. The player whose estimate is closest to the actual sum scores a point for the team. In case of a tie, both teams earn a point. The next player on each team continues the estimation game.

Subtraction rules are similar. One player from each team selects a three-digit number. Both players then write down their estimates of the difference between the two numbers. The player whose estimate is closest to the actual difference earns a point for the team. Students who engage in this activity develop estimation strategies that will benefit them in and out of school.

2. Find calculator patterns (grades 2–6). You need a calculator. Choose a number from 2 to 12. Press the + key. Press the = key. (You should see the number you first entered.). Keep pressing the = key. Each time you press, write the number displayed. Continue until there are twelve numbers on your list. Explain the patterns you noticed (Burns, 1996).

3. Calculator multiplication challenges (grades 4–9). Each student needs a calculator. You are to find the missing number by using the calculator and by using the problem-solving strategy of guessing and checking. See how many guesses each takes you. List all of your guesses. For example 4x = 87. You might start with 23 and then adjust. Below is a simple solution.

$4 \times 23 = 92$
$4 \times 22 = 88$
$4 \times 21 = 84$
$4 \times 21.5 = 86$
$4 \times 21.6 = 86.4$
$4 \times 21.7 = 86.8$
$4 \times 21.8 = 87.2$
$4 \times 21.74 = 86.96$
$4 \times 21.75 = 87$

4. Solving problems using the calculator: Find how many seconds old you are. Students may need to become familiar with the directions by first doing some research (how many seconds are there in a minute, a day, a month, a year?). How old will you be at noon today? Make a guess and write it down. Use a calculator to find out. The problem requires several steps to find a solution:

1. Decide what information is needed and where to collect it.
2. Choose the information to use.
3. Perform the necessary calculations.
4. Use judgment to interpret the results and make decisions about a possible solution.

5. Calculator counting. The calculator as a powerful counting tool can be used to teach important concepts like sequencing, placing value, and one-to-one correspondence. A child's physical interaction with this is almost a magical counting device.

To make a calculator count: enter the number 1 on your calculator and press the + sign. Press the + sign again. Then, press the = sign.

Continue to press =. The calculator will begin counting. Each time the = sign is pressed, the next number in sequence appears on the screen. If this set of directions doesn't work with your calculator, check the instructions that come with the calculator. The directions should indicate how to get a constant function. Follow the instructions on how to get a constant and any of the counting activities will work for you (Reys, Suydam, & Lindquist, 2004).

Technology is impacting all areas of our lives. The math standards suggest, in keeping with these changes, that calculators and computers can help students explore math concepts and expand mathematical understanding.

Set 2: Activities for Technology Education

Activity 1: Communications Timeline (upper grades)

Inquiry question: How have means of communications changed over time?

Concept: History influences how people communicate.

The ways in which people communicate with each other have changed throughout history. In ancient days, cave painting conveyed messages and created meaning for people. For centuries, storytelling and oral language served as the primary means of communicating information. Handwritten manuscripts were the first written for communication, followed more recently by the printing press, telegraph, typewriter, telephone, radio, television, computers, and smartphones. The list could go on.

Purpose and objectives: This is an example of a design activity that meets the math/science/technology standards. Through this activity, students will research the history of communications technology and create a timeline in their math/science journal. This activity allows students to collect as many actual objects as possible or their representations for display. They will provide a written explanation about these communications devices and talk and share ideas with others, answering any questions the class raises.

Materials: reference books, science and math journals, communication devices from home, grandparents, community, or elsewhere.

Procedure:

1. Have students conduct research on the history of communications technology and create a timeline. Have them put their notes in their math and science journal.
2. Encourage students to assemble a communications timeline project for display, using as many actual objects or their representations as possible.
3. Remind students that each time period needs to have some examples of the actual objects used and a written explanation about these communications devices.

Evaluation and extension:

1. Direct students to choose a communications technique from the past. Teachers may wish to divide students into groups according to their interests and assign each group a certain time period or technological tool used for communication.
2. Direct groups to orally (and perhaps graphically) present their communication tool to the class.
3. Teachers may extend the project by having students project what the communications scenario of the future will look like.

Activity 2: Create a Water Clock

Inquiry question: How do clocks work?

Concept: Clocks keep track of time. Time is often a difficult concept for children to grasp. People have recorded the passage of time throughout history.

Purpose and objectives: This is an example of a design activity that meets the math/science/technology standards. This activity involves children in time measurement by using a number of old and new technological tools. Students will learn how to measure time using a variety of clocks.

Materials: variety of large cans, plastic bottles, and plastic containers, a collection of corks or plugs, modeling clay, scissors or a knife, hammer and nail with a large head, math and science journal.

Procedures: Have students collect a variety of large cans, plastic bottles, and plastic containers. You may wish to help them make a small hole in the bottom of the metal containers with a hammer and large nail or in the plastic containers using a scissors or a knife (try to make all of the holes in the containers the same size). Instruct students to make a clay plug or a small cork to fit the hole. Have students fill the containers with water, then release the plugs and compare the time it takes for each container to empty. Encourage students to guess which one will empty first.

Evaluation and follow-up:

1. Have students choose common jobs that can be timed with water clocks.
2. Encourage students to make a list of things that can be timed with a water clock.
3. Instruct students to hypothesize what the effect of different sized holes is on the water drip process.
4. Have students use a digital clock to determine how much water flows out in one minute's time from their water clock.
5. Ask students to design a system to mark their water clock to determine the time without measuring the water each time.
6. Ask students if they can make a clock another way.
7. Have students write a program for a computer to record time.

Follow-up questions: Instruct students to respond to these questions in their math and science journal:

- Why are clocks so important to the industrial age?
- How are clocks used as metaphors?
- Encourage students to speculate on the future of clocks and their role in the future.

Activity 3: Hypothesis Testing

Inquiry question: How do I find out?

Concept: This technology awareness activity is designed to get students involved in the historic role of technology in today's society.

Purpose and objectives: Students will conduct inquiry in trying to discover what technological devices are being presented. Students will reinforce their skills of questioning, observing, communicating, and making inferences. This is an example of a design activity that meets the math/science/technology standards.

Materials: Instruct students to bring in a paper bag containing one item that no one would be able to recognize (an old tool of their grandfather's, for example), one item that some people may be able to identify, and one common item that everyone would recognize.

Procedure:

1. Divide students into small groups. Tell students that all items in their bags should be kept secret.
2. Give the students the following directions:
   a. There will be no talking in the first part of this activity.
   b. You are to exchange bags with someone else in your group.
   c. You may then open the bag, remove one item, and write down what you think that item is. (Have students examine each item carefully. Also, have students write their reaction to how they feel about this item, what they think it may be used for, and which category this item falls into: common item, one no one would recognize, or an item some will recognize).
3. Repeat this process with each of the items in your bag.
4. Exchange bags with other groups and go through the same procedure.

Evaluation, completion, and follow-up: When all students have finished examining their bag of articles and written their responses, they should meet back in their groups and explain what they have discovered in their bags. Encourage class speculation, questions, and guesses about unidentified items. The student who brought in the unknown tool or article should be responsible for answering the questions posed but should not identify the item until all guesses and hypotheses have been raised.

When students engage in formal and informal math, science, or technology activities, they develop and use a number of science process skills. From the early childhood years onward, children use the skills of observing, classifying, measuring, estimating, inferring, predicting, and communicating. It is best to develop these skills in the context of social interaction; so it's best to pay attention to teamwork skills during scientific inquiry and mathematical problem solving. To encourage creative and innovative thinking, teams need to be small, dynamic, and flexible—and they also need to stay focused on the content of the lesson.

Suggestions for teachers and parents:

- Praise effort more than achievement.
- Teach delayed gratification.
- Limit reprimands.
- Use praise to stimulate curiosity.

At home or at school, a bad environment suppresses creative and innovative development.

Set 3: Activities for Optimizing Multimedia Options

Finding Video Clips Online

Using multimedia is a good way to activate prior knowledge and differentiate mental models in a way that helps a wide range of students acquire new knowledge.

Some suggested resources:

- United Streaming (streaming.discoveryeducation.com). This is a good educational video collection and a good advance organizer for starting a learning activity.
- Google Videos (video.google.com). This search engine looks for video clips using the keywords you enter.
- Creative Commons (creativecommons.org). This is a nonprofit engine that searches flexible copyright use for anyone putting together a creative package that requires video, graphics, sounds, or publications.
- A9 (a9.com). This search engine from Amazon.com searches for images, movies, blogs, books, and other websites.
- The Internet Archive (www.archive.org). This resource has video clips from the entire twentieth century. A feature called the Way Back Machine is particularly popular with students.

By creating original multimedia stories, students can't help but see the inner processes of media. The new concept of literacy is bound to involve "reading" and "writing" in multiple media formats, connecting to the curriculum, and integrating the results into a meaningful whole.

Many teachers who are unfamiliar with multimedia technology overestimate the difficulty of getting students started. Just put them with a partner and ask them to play with a couple of sites first. The next step is figuring out an interesting project. Another approach: have students explore these sites as homework; when they come to class, they can team up and explore where to go on in a way that connects to the curriculum.

MAKING SENSE OF ELECTRONIC
IMAGERY AND OPTIMIZING MULTIMEDIA OPTIONS

Learn to Critique Content

Decoding visual stimuli and learning from visual images require practice. Seeing an image does not automatically ensure learning from it. Students must be guided in decoding and looking critically at what they view. One technique is to have students "read" the image on various levels. Students identify individual elements and classify them into various categories, then relate the whole to their own experiences, drawing inferences and creating new conceptualizations from what they have learned. Encourage students to look at the plot and story line. Identify the message of the program. What symbols (camera techniques, motion sequences, setting, lighting, etc.) does the program use to convey its message? What does the director do to arouse audience emotion and participation in the story? What metaphors and symbols are used? (These activities are examples of a design activity that meets the math/science/technology standards.)

Compare Print and Video Messages

Have students follow a current event on the television evening news and compare it to the same event as written about in a major newspaper. A question for discussion may be, how do the major newspapers influence what appears on a national network's news program? Encourage comparisons between both media. What are the strengths and weaknesses of each? What are the reasons behind the different presentations of a similar event?

Evaluate Viewing Habits

After compiling a list of their favorite programs, assign students to analyze the reasons for their popularity and examine the messages these programs send to their audience. Do the same for favorite books, magazines, newspapers, films, songs, and computer programs. Look for similarities and differences between the media.

Use Video for Instruction

Use a DVD player or computer projection equipment to make frequent use of three- to five-minute video segments to illustrate different points. This is usually better than showing long segments. For example, teachers can show a five-minute segment from a movie to illustrate how one scene uses foreshadowing or music to set up the next scene.

Analyze Advertising Messages

Advertisements provide a wealth of examples for illustrating media messages. Move students progressively from advertisements in print to television commercials, allowing them to locate features (such as packaging, color, and images) that influence consumers and often distort reality. Analyze and discuss commercials in children's programs: How many minutes do TV or Internet ads appear in an hour? How have toy or game manufacturers exploited the medium? What is the broadcaster's or developer's role? What should be done about it?

Create a Scrapbook of Media Clippings

Have students keep a scrapbook of newspaper and magazine clippings on electronic media. Paraphrase, draw a picture, or map out a personal interpretation of the articles. Share these with other students.

Create New Images from the Old

Have students take rather mundane photographs and multiply the image, or combine them with others, in a way that makes them interesting. The artist David Hockney calls these "Joiners." Through the act of observing, it is possible to build a common body of experiences, humor, feeling, and originality. And through collaborative efforts, students can expand on ideas and make the group process come alive.

Use Debate for Critical Thought

Debating is a communications model that can serve as a lively facilitator for concept building. Taking a current and relevant topic and formally debating it can serve as an important speech/language extension. For example, the class can discuss how mass media can support political tyranny, public conformity, or the technological enslavement of society. The discussion can serve as a blend of math, science, technology, and humanities studies. You can also build the process of writing or videotaping from the brainstorming stage to the final production.

Include Newspapers, Magazines, Literature, and Electronic Media (like Brief Television News Clips) in Daily Class Activities

Use of the media and literature can enliven classroom discussion of current conflicts and dilemmas. Neither squeamish nor politically correct, these sources of information provide readers with something to think and talk about. And they can present the key conflicts and dilemmas of our time in a way that allows students to enter the discussion.

These stimulating sources of information can help the teacher structure lessons that go beyond facts to stimulate reading, critical thinking, and thoughtful discussion. By not concealing adult disagreements, everyone can take responsibility for promoting understanding by engaging others in moral reflection and providing a coherence and focus that help turn controversies into advantageous educational experiences.

With today's interactive multimedia programs, there is every reason to expect math, science, and technology education programs that can invite students to interact with creatures and phenomena from the biological and physical universe. Students can move from the past to the future and actively inquire about everything from experiments with dangerous substances to simulated interaction with long dead scientists. Just don't leave out experiments with real chemicals and experiences with live human beings.

Good educational software often tracks individual progress over time and gives special attention to problem areas. Also, there is a tendency to move away from the computer platform and put educational programming on all kinds of gadgets. Even *Children's Software Review* has changed its name to *Children's Technology Review*. One of their links, www.littleclickers.com, is a good site for finding educational games available on the Internet.

Good things can happen when students and teachers use computer-based technology and information networks. For example:

- Computer-based simulations and laboratories can be downloaded in support of the standards.
- Students can more easily become involved in active participatory learning.
- Networking technology, like the Internet, can help bring schools and homes closer together.
- Technology and telecommunications can help include students with a wide range of disabilities in regular classrooms.
- Distance learning, through networks like the Internet, can extend the learning community beyond the classroom walls.
- The Internet may help teachers continue to learn—while sharing problems/solutions with colleagues around the world.

In classrooms where computers are used, one of the first things that teachers learn is when to say "computers off"—or "screens down" if learners are using laptops. Remember, anything that diminishes attention and rapport with others gets in the way of deep learning.

Since the Internet is rarely censored, it is important to supervise student work or use a program that blocks the most objectionable websites. We suggest that teachers keep an eye on what students are doing and make sure that the classroom is offline when a substitute teacher is in. A program like Net Nanny is one way to prevent children from accessing inappropriate material. Just as with libraries and bookstores, it is important not to restrict the free flow of information and ideas. But there can be a children's section without bringing all content down to an intellectual level appropriate for a seven-year-old.

In today's world, many children grow up interacting with electronic media as much as they do print or people. At school, digital technology makes it relatively easy to engage even the most reluctant learner. But does this mean that students are learning anything meaningful or that they are making good use of either educational *or* leisure time?

The Internet, like other electronic media, can distract students from face-to-face interaction with peers—inhibiting important group learning functions and physical exercise activities. Still, although the future may be bumpy, it doesn't have to be gloomy. Good use of any learning tool depends on the strength and capacity of teachers. The best results occur when informed educators are making instructional decisions, rather than simply following a path set by the technology itself.

Harmonizing the present and an even more technologically intensive future has a lot to do with improving the schools. Digital tools have great promise, but anyone who thinks that technological approaches will solve most educational problems is mistaken. It's important to turn the volume down on the cacophony of conflicting directives and specious conjectures about the educational future and focus on what is known about instruction. Research frequently points to what works—including technology, standards, textbooks, activities, and assessment. But when it comes to actually doing what works, the most important factor affecting how much students learn is teachers.

COLLABORATION, DIFFERENTIATION, AND IMAGINATIVE ENGAGEMENT

After discussions in small collaborative groups, projects or work assignments can be brought back to the whole class and each group shares their findings. Sometimes, a group may want to put their findings online or post their book reviews at www.amazon.com. Wikipedia has open editing, so students can put some things there. This online encyclopedia is very timely, but the accuracy is mixed. So we tell students that it's a good starting point but that it shouldn't be their only source.

When students are actively engaged with ideas and other students, the natural power of teamwork accommodates more types of learning than the old chalk and teacher talk model. It has always been true that when interesting questions are raised in learning groups, those involved tend to lead each other forward.

Sometimes students need to take conscious steps to activate prior knowledge. This can be done as a small group reviews what's been covered verbally and on paper. Collaborative learning of this type is effective because the framework of the strategy is good for all students. The research also suggests that somewhat-collaborative learning groups result in more cross-cultural friendships and have some positive effect on intergroup relations. With an increasingly diverse group of students, learning to advance through the intersection of different points of view is more important than ever.

While aiming high, teachers have to be realistic about what children and young adults can achieve. To help all students, teachers need to focus on the concepts that they want learners to deal with. The next step is figuring out how different kinds of learners are going to use different methods to show an understanding of what's covered. Digital technology can help students focus more on certain aspects of a subject that are of most interest to them.

Differentiation does not mean that teachers have to create separate tasks for every student in the classroom. Differentiated instruction, though, is a natural step to providing students with a certain level of control over content. Better yet, this approach can encourage learners to jointly construct meaning and understanding by interacting with others.

DI is not individualization; it's an approach that encourages looking for patterns of needs and interests. It opens doors to planning in a way that meets the multiple needs that learners bring to the classroom. Also, students might work in pairs or in small groups to express what they have learned in several possible ways. With different groups taking different paths to content mastery, more than one thing is happening in the classroom. Additionally, different media may be used to communicate the results in different ways.

Digital media has great promise for addressing varied needs thoughtfully and smoothly. All the while, students can be creatively engaged in learning. It is important that teachers provide extra enrichment for their high-achieving students so that they stay challenged and their parents stay cooperative. Contrary to public opinion, intelligent people are not always the most innovative thinkers; for one thing, they may be tempted to use their abilities to support a particular position rather than constructively exploring new questions and problems.

Creating a more innovative classroom environment requires paying attention to developing the innovative and creative potential that all children possess. Innovative people vary in their natural ability levels, motivation, and interests. However, they may share certain characteristics. They tend to be

- flexible in their thinking
- curious and resourceful
- able to solve difficult problems
- comfortable outside traditional boundaries
- good at seeing new implications
- able to synthesize different possibilities
- self-confident and willing to take risks

Sparking innovative insights requires everyone from teachers to scientists to find ways of inspiring young people to create and innovate so that they can become inventors and makers of things. It is important to be more than just consumers of what someone else developed.

After doing a unit of study, ask those involved to make two columns and reflect on the following.

I used to think: *I now think:*

With a partner, discuss any changes (if any) that have taken place. This activity works for students—and for teachers in a professional development workshop or class.

You don't have to be a computer technician to lead students into a digital-intensive future. However, you do have to know how to orchestrate learning conditions in a way that brings out the best in everybody by effectively managing students' time, talents, and productivity. This means implementing a variety of instructional and grouping strategies in a way that meets the diverse needs of learners. In addition to all this, teachers are often called upon to make appropriate choices about technology systems, resources, and services.

There is wide agreement that new technologies should be used in ways that closely correspond with real-world uses in order to prepare students for further education and the workplace. Technological fluency is more than a technical skill set; it involves weaving creativity and innovation into lessons, assignments, and students' lives in ways that connect to subject matter. Educators, parents, the media, and society, in general, are all responsible for developing the next generation of innovators.

Besides altering how we learn, play, live, and work, technology has become a powerful tool for sparking students' imagination and helping teachers teach for understanding. It can even help puncture some of the colorful balloons of mathematical nonsense, pseudoscience, and technotrash. And, yes, computers and the Internet give students access to more people and more information. But digital tools are still "works in progress"—with changes and improvements happening at a rapid rate. So it's important that enthusiasm not trump judgment; schools don't need expensive distractions from core instructional responsibilities.

If faith in technology simply becomes a powerful ideology, we miss an important point: technology is an important thing, but not the only thing. It can be magical, but it is not the main purpose in life and it is not a silver bullet for educational improvement. When it comes to new approaches to learning with (and about) technology, a little skepticism will improve the product. Remember, although the chalkboard worked out, technological shortcuts from filmstrips to videotapes have promised a lot and delivered a lot less. Digital technology promises much more.

When it comes to motivating the innovators of the future, it requires the right mix of policy, leadership, funding, and culture. It's important to remember that when it comes to technology, thinking about the educational process has to come first.

It is always a challenge to reorient values and priorities in a way that leads from rhetoric to a shared sense of moving toward a better reality. Remember, people have a tendency to relabel their existing practices with whatever ideas are the current fashion. So the journey is smoother if you play less attention to what people say they believe and more attention to what you observe them doing.

As far as quality of instruction is concerned, much depends on daring to explore the outer limits of your knowledge with self-confidence, enthusiasm, and a clear understanding of the characteristics of effective instruction.

Before teachers can differentiate technology instruction, they have to know how to modify content, processes, and products in a way that reflects student differences. But even if teachers do a good job of filling in some of the educational potholes, a world-class school system requires a sustained societal commitment to developing an agile, smarter, and technologically savvy citizenry.

THE INTERSECTION OF CULTURE, EDUCATION, AND TECHNOLOGY

The intersection of culture, education, and technology has always been an unsettled place driven by human creativity. From the clay tablets to the printing press and on to computers, important technologies have changed our sense of time and space. Each new medium altered how we access information, communicate, and learn. An important difference: new digital devices seem to be taking more control than the technology of the past.

Concerns about disappearing into the consensual hallucination of cyberspace have not slowed down the desire to surf the Net. And the last ten years have brought us some really good search possibilities, smartphones, and social networking. The next ten years are likely to bring us digital tools to monitor our health, more advanced robots, and digital tools that can create the most complex imagery using a cheap computer.

Today's Internet has certainly been transformative. But it's not always a powerful force for good. In their day of media dominance, printed text, the telegraph, radio, and television were all seen as world-changing technologies. They all made a significant difference, but none have quite lived up to the expectations of their most ardent supporters. The web seems to be following the same pattern: a combination of good, bad, and even ugly. There are, after all, ways that transnational networks (fueled by digital technologies) worsen the world as we know it.

There is a lot to be frightened of. But if anyone thinks digital technology is going away anytime soon they will be sadly disappointed. There is a lot to be excited about in today's media revolution. Take the Internet as an example—it can be a useful two-way medium. In fact, the collaborative possibilities found on the web provide some very interesting possibilities for education. The key is making sure that technology not be used to replace or reduce the role of the human teacher. The overall educational goal: to serve specific academic purposes and facilitate the creation of community and a truly human society.

As far as the schools are concerned, technological usefulness has a lot to do with enabling learners. For this to happen, digital approaches have to enhance what the teacher is trying to do and avoid being an annoying distraction. This means that there are times when we have to challenge the cult of speed and turn off the intrusive array of digital devices.

Like many tech companies, we encourage everybody to stay away from e-mail and electronic gadgets for at least one workday per week. Still, in the hands of thoughtful and informed teachers digital technology can be a powerful lever for amplifying creativity and motivating students with a wide range of social backgrounds and individual needs.

To some extent, everyone is innovative, but there are considerable differences when it comes to the environmental and personal possibilities for imaginative expression. A culture that fosters self-expression is much more likely to be at the center of human creativity in the arts and sciences. If you mix inventive tendencies with the ability to deal with disruptive change, it gives a powerful advantage to both individuals and countries.

As far as the classroom is concerned, there should be a greater emphasis on developing *every* child's imaginative potential. Digital technology can help in the classroom by providing possibilities for collaborative activities, creative engagement, and thoughtful invention.

Computers and other digital tools can assist the differentiation of instruction—and this helps make classes more relevant, engaging, collaborative, and thoughtful. They can also provide teachers with access to a wide range of instructional tools and better communication with parents. Technology can also contribute to the process of learning how to learn—an important skill for school and life in the twenty-first century. Understanding technology and being able to learn new things quickly matters, whether you go to school for two or ten years after high school.

CONCLUSION: SPARKING INNOVATION IN THE TWENTY-FIRST CENTURY

One can think of ordered arenas of change as an inverted pyramid:

Large-scale changes involving the diverse population of a region or an entire nation
Large-scale changes involving a more uniform or homogeneous group
Changes brought about by science, technology, or the arts
Changes within formal instructional settings
Intimate forms of mind changing
Changing one's own mind

(Adapted from a concept presented by Howard Gardner, 2004)

It is a social responsibility to make sure that every student is educated to his or her full potential. For some, university education is the way to go; for others, a community college might be better. Whatever path they take, all K–12 students need to become college or career ready. While everyone would benefit from some postsecondary education, everybody would not enjoy a full-fledged university program.

Sometimes you need a philosopher, and sometimes you need a plumber. Skilled trades such as those of an electrician, carpenter, mechanic, and plumber cannot be outsourced—and the pay is better than substitute teaching or being an adjunct professor. Germany and Japan are but two examples of countries that have great apprenticeship programs for those who are not inclined toward a life of the mind.

As we have conversations about the kind of education every person deserves, it is important to remember that the classroom of the future will be even more socially and educationally diverse than it is today. Whatever their background or interests, all students will need to learn how to use digital media individually and collectively. Such a level of technological competency involves being able to work with techno-logical tools—and using those same tools to gain an understanding of technology.

As students pass through the cauldron of technological change, their creative efforts can be reinforced by formative assessment, feedback, and recognition. Many schools have long encouraged *some* students to think creatively, reason, problem solve, and use technology. What's new is that, now, there is an effort to involve all students in all schools to take part in such higher-level activities.

A useful slogan for fostering innovation: develop a healthy disregard for the impossible.

It is important to be able to find and focus on significant problems. This is increasingly difficult in a multimedia age where information is con-stantly distracting us in a way that shortens everybody's attention span.

Imaginative people are also good at imagining what could be—and they don't let perceived limits stop them from doing something. Tenac-ity matters as much today as it did in the past. As we travel along the path to the future, you can be certain that whatever unexpected events or the forces of invention bring us, fostering and accelerating innova-tive behavior will remain an educational constant.

The crucial mechanisms for determining both individual and national success in the twenty-first century are innovation, knowledge, and an understanding of technology. There is a lot more scope for computers so you can be sure that the digital revolution has more surprises in store for us. Since computers are nowhere near the end of their logical development, there is still plenty of room for digital devices making our lives more interesting.

In the next few decades, we will cross some surprising technology-related thresholds. Forecasting the future is a little like predicting earth-quakes. A lot of people are trying to figure it out, but accurate predictions remain elusive. For now, the best approach is to spend less time worrying about the unpredictable and more time preparing for the inevitable.

REFERENCES

Adams. D. & Hamm, M. (2001). "Literacy, Learning, and Media." In Ohler, J. (Ed.), *Future Courses: A Compendium of Thought about Education,*

Technology, and the Future. Bloomington, IN: Press of the Agency for Instructional Technology, pp. 43–56.

Anderson, C., & Wolff, M. (2010, August 17). The web is dead. Long live the Internet. *Wired Magazine.* Retrieved August 15, 2010, from www.wired.com/magazine/2010/08/ff_webrip/all/1.

Burns, M. (1996). *50 Problem-Solving Lessons: The Best from 10 Years of Math Solutions Newsletters.* Sausalito, CA: Math Solutions.

Cathcart, G., Pothier, Y. M., Vance, J. H., & Bezuk, N. S. (2007). *Learning Math in Elementary and Middle School & IMAP Value Package.* Upper Saddle River, NJ: Allyn & Bacon.

Gardner, H. (2004). *Changing Minds: The Art and Science of Changing Our Own and Other People's Minds.* Cambridge, MA: Harvard Business Press.

International Technology Education Association (ITEA). (2000). *Standards for Technological Literacy: Content for the Study of Technology.* Reston, VA: ITEA.

Kaiser Family Foundation. (2010). *Daily Media Use among Children and Teens Up Dramatically from Five Years Ago.* Menlo Park, CA: Kaiser Family Foundation.

Kirkpatrick, D. (2010). *The Facebook Effect: The Inside Story of the Company That Is Connecting the World.* New York, NY: Simon & Schuster.

National Research Council (NRC). (2001). Classroom Assessment and the National Science Education Standards 2001. Washington, DC: National Academy Press.

National Research Council (NRC). (2000). Education Teachers for Science, Mathematics, and Technology: New Practices for the New Millennium 2000. Washington, DC: National Academy Press.

Reys, R. E., Suydam, M. N., & Lindquist, M. M. (2004). *Helping Children Learn Mathematics*, 7th ed. Hoboken, NJ: John Wiley & Sons.

Turkle, S. (2011). *Alone Together: Why We Expect More from Technology and Less from Each Other.* New York, NY: Basic Books.

RESOURCES

Abramovich, S. (2010). *Topics in Mathematics for Elementary Teachers: A Technology-Enhanced Experiential Approach.* Charlotte, NC: Information Age Publishing.

Bell, G., & Gemmell, J. (2009). *Total Recall: How the E-Memory Revolution Will Change Everything.* New York: Dutton.

Benjamin, A. (2005). *Differentiated Instruction Using Technology: A Guide for Middle and High School Teachers*. Larchmont, NY: Eye on Education.

Bern, B., & Sandler, J. (2009). *Making Science Curriculum Matter: Wisdom for the Reform Road Ahead*. Thousand Oaks, CA; Corwin Press.

Bers, M. (2008). *Blocks to Robots: Learning with Technology in the Early Childhood Classroo*m. New York: Teachers College Press.

Bouerlein, M. (2008). *The Dumbest Generation: How the Digital Age Stupefies Young Americans and Jeopardizes Our Future*. New York: Penguin.

Brown, H. (2007). *Knowledge and Innovation: A Comparative Study of USA, the UK, and Japan*. New York: Routledge.

Brown, S. (2009, March 23). On stand-up comedy, lingua franca of the wired world. *Wired Magazine*.

Butz, S. (2010). *Computer Projects Grades 5–6*. Westminster, CA: Teacher Created Resources.

Carr, N. (2010). *The Shallows: What the Internet Is Doing to Our Brains*. New York: W.W. Norton & Company. [This book explores how increasing Internet use is diminishing our ability to think deeply and creatively.]

Dehaene, S. (2009). *Reading in the Brain: The Science and Evolution of a Human Invention*. New York: Viking Press.

Dyson, E. (2010). *Release 2.0*. New York: Broadway. [This book focuses on how computers and the Internet can empower users.]

Hamilton, B. (2007). *It's Elementary! Integrating Technology in the Primary Grades*. Eugene, OR: International Society for Technology in Education.

International Society for Technology in Education (ISTE). (2000). *National Educational Technology Standards*. Washington, DC: ISTE. [In 2000, the ISTE revised its previously published standards for teachers, ISTE Technology Standards for All Teachers, and released it as NETS for Teachers, or NETS•T.]

International Technology Education Association (ITEA). (2004). *Advancing in Technological Literacy; Assessment, Professional Development, and Program Standards*. Reston, VA: ITEA.

Keen, A. (2008). *The Cult of the Amateur: How Blogs, MySpace, YouTube, and the Rest of Today's User-Generated Media Are Destroying Our Economy, Our Culture, and Our Values*. New York: Broadway Business.

Manjoo, F. (2008). *True Enough: Learning to Live in a Post-Fact Society*. Hoboken, NJ: John Wiley & Sons. [This book explores how new technologies tend to foster personal belief over facts.]

Mayer-Schonberger, V. (2009). *Delete: The Virtue of Forgetting in the Digital Age*. Princeton, NJ: Princeton University Press.

Menn, J. (2010). *Fatal System Error: The Hunt for the New Crime Lords Who Are Bringing Down the Internet*. New York: Public Affairs. [This book examines how cybercriminals use the Internet to reach across borders.]

Mitchell, M. (2010). *Elementary Mathematics through Technology: Integrating Technology, Mathematical Content and Pedagogy*. New York: Routledge.

Napier, H. A., Rivers, O. N., & Hoggatt, J. P. (2011). *Learning with Computers I (Level Green Grade 7)*. Cincinnati, OH: South-Western Educational Publishing.

————. (2011). *Learning with Computers I (Level Orange, Grade 8)*. Cincinnati, OH: South-Western Educational Publishing.

Negroponte, N. (1996). *Being Digital*. New York: Vintage Books. [This book takes a techo-utopian view of digital technology.]

Ohler, J. (2007). *Digital Storytelling in the Classroom: New Media Pathways to Literacy, Learning, and Creativity*. Thousand Oaks, CA: Corwin Press.

Page-Botelho, M. (2010). *Teaching Lower Elementary Technology*. Page-Botelho.

Parker, J. K. (2010). *Teaching Tech-Savvy Kids: Bringing Digital Media into the Classroom, Grades 5–12*. Thousand Oaks, CA: Corwin Press.

Parker, L. (Ed.). (2008). *Technology-Mediated Learning Environments for Young English Learners: Connections In and Out of School*. New York: Lawrence Erlbaum Associates.

Plowman, L., Stephen, C., & McPake, J. (2010). *Growing Up with Technology: Young Children Learning in a Digital World* (Kindle Edition). London: Taylor and Francis.

Rosen, L. (2010). *Rewired: Understanding the iGeneration and the Way They Learn*. New York: Macmillan.

Strickland, C. (2009). *Exploring Differentiated Instruction*. Alexandria, VA: Association for Supervision and Curriculum Development.

Wenglinsky, H. (2005). *Using Technology Wisely: The Keys to Success in Schools*. New York: Teachers College Press.

Yelland, N. (2007). *Shift to the Future: Rethinking Learning with New Technologies in Education*. New York: Routledge.

Zucker, A. (2008). *Transforming Schools with Technology: How Smart Use of Digital Tools Helps Achieve Six Key Education Goals*. Cambridge, MA: Harvard Education Press.

www.ingramcontent.com/pod-product-compliance
Lightning Source LLC
Chambersburg PA
CBHW021818270326
41932CB00007B/246